Dialogues Around Models and Uncertainty
An Interdisciplinary Perspective

Dialogues Around Models and Uncertainty

An Interdisciplinary Perspective

editor

Pauline Barrieu

London School of Economics and Political Science, UK

NEW JERSEY · LONDON · SINGAPORE · BEIJING · SHANGHAI · HONG KONG · TAIPEI · CHENNAI · TOKYO

Published by

World Scientific Publishing Europe Ltd.
57 Shelton Street, Covent Garden, London WC2H 9HE
Head office: 5 Toh Tuck Link, Singapore 596224
USA office: 27 Warren Street, Suite 401-402, Hackensack, NJ 07601

Library of Congress Cataloging-in-Publication Data
Names: Barrieu, Pauline, editor.
Title: Dialogues around models and uncertainty : an interdisciplinary perspective /
 editor, Pauline Barrieu, London School of Economics and Political Science, UK.
Description: New Jersey : World Scientific, [2020] | Includes bibliographical references.
Identifiers: LCCN 2019035976 | ISBN 9781786347749 (hardcover) |
 ISBN 9781786347756 (ebook) | ISBN 9781786347763 (ebook other)
Subjects: LCSH: Statistics--Methodology. | Uncertainty--Mathematical models. |
 Statisticians--Interviews.
Classification: LCC QA276.A2 D53 2020 | DDC 519.5/4--dc23
LC record available at https://lccn.loc.gov/2019035976

British Library Cataloguing-in-Publication Data
A catalogue record for this book is available from the British Library.

Copyright © 2020 by World Scientific Publishing Europe Ltd.

All rights reserved. This book, or parts thereof, may not be reproduced in any form or by any means, electronic or mechanical, including photocopying, recording or any information storage and retrieval system now known or to be invented, without written permission from the Publisher.

For photocopying of material in this volume, please pay a copying fee through the Copyright Clearance Center, Inc., 222 Rosewood Drive, Danvers, MA 01923, USA. In this case permission to photocopy is not required from the publisher.

For any available supplementary material, please visit
https://www.worldscientific.com/worldscibooks/10.1142/Q0230#t=suppl

Desk Editors: Vishnu Mohan/Jennifer Brough/Shi Ying Koe

Typeset by Stallion Press
Email: enquiries@stallionpress.com

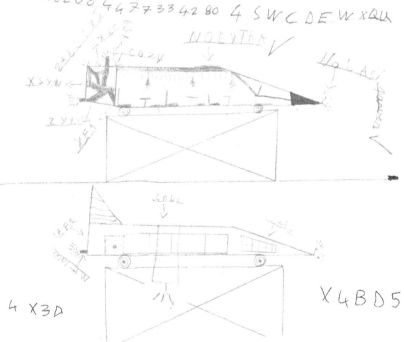

Models — Arthur Albertini (2018)

Preface

This book project has been with me for many years, probably even longer than I can think of. It is really a personal project or, I would even dare to say, a personal quest to understand better the fundamentals of research and modelling.

Being such a personal adventure, let me start by mentioning briefly where it probably all started. My family is an artistic family. My dad is a pure researcher. He has always been. Research in Art can be extremely fundamental, focusing on the essence of things itself, "exactitude" versus "vérité," using the distinction of Henri Matisse. Or quoting one of the books which has been very important in my life, *Le chef d'œuvre inconnu* from Balzac, where Frenhofer says "[…] l'ombre palpitante des cils flotte ainsi sur les joues! C'est cela, et ce n'est pas cela. Qu'y manque-t'il? Un rien, mais ce rien est tout. Vous avez l'apparence de la vie, mais vous n'exprimez pas son trop-plein qui déborde, ce je ne sais quoi qui est l'âme peut-être et qui flotte nuageusement sur l'enveloppe […]."[1]

As far as I can remember, deep discussions about things, about the cause of things, were going on. Dinners with painters, musicians, and

---

[1] "[…] the quivering shadow of the lashes hovers upon her cheeks. It is all there, and yet it is not there. What is lacking? A nothing, but that nothing is everything. There you have the semblance of life, but you do not express its fullness and effluence, that indescribable something, perhaps the soul itself, that envelopes the outlines of the body like a haze […]" (*CreateSpace Independent Publishing Platform*, 2012).

philosophers were frequent and from a young age I was allowed to sit at the table and listen to these exchanges. Not surprisingly, all the jobs I wanted to do as a young girl involved some research component, from archaeologist to forensic doctor!

In my studies, even before contemplating the idea of a PhD, I was always driven by multiple forces and was always reluctant to specialise in a particular discipline: literature and philosophy, mathematics and science… I simply could not choose. For the doctoral studies, it was the same…and so life chose for me and I was very lucky to be allowed to adventure on two different fronts and worked on two PhDs, one in applied mathematics and one in finance, seeing this as part of the same journey.

Exciting areas of research have been for me, since the very beginning, at the fringe, at the border, at the junction of different fields, emerging from the encounter with another researcher. I have been extremely influenced by my PhD supervisor in mathematics, Nicole El Karoui, in my approach to research. She showed me and shared with me her vision of research as a human experience, as an exchange, as a generous moment. From very early discussions with researchers from other areas, beyond the excitement of the discovery of relevant questions, another aspect started to emerge: the importance of understanding the other, of explaining the basics, what can be seen as the convention of our respective "fields," anything that one takes for granted, the foundations, the "culture."

It appeared to me that, maybe, in these differences of methods, of approaches, of processes, of convention, something deeper was present. Maybe by questioning the fundamentals and dissecting them, we could understand better how our minds work, and would be able to think differently when considering various questions in our "fields." Hence, this project! I remember a book I read during my doctoral studies, *Dialogues autour de la création mathématique* by Nicolas Bouleau (Association Laplace–Gauss, 1997). This book has been inspiring in many different ways, especially in its aim at clarifying and understanding the "obvious," but also communicating the passion of all the mathematicians involved with the book. It has had a big impact on me, because of its approach, its questioning.

I started this book project with the aim of questioning the fundamentals, I didn't expect, for a second, to enjoy so much the project itself. Each interview, each meeting has been a moment of extreme generosity, openness, and sharing. The energy and stimulation each contributor has given me have been incredible. I hope some of it, if not all, will transpire from the various contributions.

Each contributor was asked the same set of questions and asked to reply in a way that is accessible to a non-expert audience. Their respective answers for each question are very different, more than I thought it would be, underlying even more not only the importance of clarification when undertaking interdisciplinary research, but also the possibility of discovering a new world just next door. All the contributors are, however, united in their passion.

Before starting the interviews, I had a plan, a structure of the book in mind with a section synthesising the "results." Having all these contributions, these gifts from extraordinary people, I feel that my role has changed. I am honoured to have been able to collect these thoughts and truly believe that my role is now just to act as a facilitator, as a transmitter. I hope this collection will generate more interest in interdisciplinary questioning and collaboration. In these days, when research is quantified by publications and normalised by subject classification, I am simply hoping for more discovery of the other disciplines, the questioning of the usual and standard benchmarks and milestones, of the fundamentals, for the simple pleasure of exploring, sharing ideas, and "doing research."

In conclusion, the original aim of this book has been to develop a better understanding of how researchers from different scientific backgrounds view models and uncertainty, how modelling works in their fields. This would represent a key step in fostering and encouraging interdisciplinary research, which seem to be essential in addressing a number of big issues society faces today, such as climate change, longevity, risk management, etc. This book consists of a collection of thoughts and views, understandings from eminent experts in their respective fields, on the topic of modelling and uncertainty. What has come up during the process of conducting these interviews and meeting the contributors goes far beyond the original aim. This is about the care, and the generosity, the passion of these eminent researchers.

Last but not least, I would like to thank Lyn Grove for her considerable help in the transcription of the various interviews, Luca Albertini and Leadenhall Capital Partners for their support, without which this project would not have been possible, and of course, all the contributors for their time and generosity.

About the Editor

Pauline Barrieu is Professor of Statistics at the London School of Economics and Political Science. She joined the Department of Statistics in 2002 having obtained two doctorates: a PhD in Finance with highest honours, H.E.C. Graduate Business School, France, awarded in October 2002; and a PhD in Applied Mathematics with highest honours, Laboratoire de Probabilités et Modéles aléatoires, University of Paris VI, France, awarded in December 2002. She received the Prize for the best actuarial PhD dissertation, Prix Actuariat, in 2003. She was recently awarded the Prix Louis Bachelier in 2018 in recognition of her contributions covering a wide range of topics spanning stochastic analysis, financial mathematics, quantitative finance, risk management, insurance, and environmental economics. Pauline has produced fundamental, influential works in financial and actuarial mathematics, stochastic analysis and a wide range of applications. Her research interests include model uncertainty, insurance-linked securitization, contract designing, micro-insurance, weather derivatives and environmental economics.

About the Contributors

Ching-to Albert Ma is Professor of Economics at Boston University. He was Associate Chair of the Boston University Economics Department between 2010 and 2016, and held a visiting appointment at the Economics Department, University of Oslo from 2005 to 2017. Albert Ma's research areas are in industrial organisation, incentives, and health economics. His recent papers study physician agency and asymmetric information, product differentiation, quality reports and incentives, and experts' incentives in the market and in an organisation. He has lectured on industrial organisation and health economics at universities in Finland, Germany, Italy, Norway, Spain, and Switzerland. During academic year 2009–2010, he was *Cátedras de Excelencia* at Universidad Carlos III de Madrid. Between 1996 and 2006, he was a co-editor of the *Journal of Economics & Management Strategy*. From 2006 to 2009, he was an editor of the *BE Journal of Economic Analysis & Policy*, and from 2004 to 2011 an associate editor of the *Rand Journal of Economics*. In 1998, together with co-author Thomas McGuire, Albert Ma was awarded the Kenneth J. Arrow Award in Health Economics. He received a PhD from the London School of Economics in 1988. Before that, he studied at the University of Manchester and at the University of Hong Kong.

Aldo Zollo is Full Professor of Seismology and Digital Signal Processing at the University of Naples "Federico II," obtaining a physics degree there in 1983, and defending his PhD thesis in Solid Earth Geophysics (Seismology) at the University of Paris VII in 1990. In his research, he deals with theoretical and experimental aspects of seismic wave propagation and fracture processes of the Earth's crust that occur during earthquakes. He has been responsible for several research projects funded by national and international agencies in the areas of seismic exploration of volcanoes, the seismic source process modelling, and the development of early warning systems for earthquake prevention.

He has been a member of the National Group of Experts for the Evaluation of the Research for the "Earth Sciences" area (GEV04) in Italy (2011–2012), a member of the National Commission for Forecasting and Preventing Great Risks nominated by the President of the Council of Ministers (2012–2015), a member of Scientific Councils of INGV (Italy) (2001–2010) and ISTerre (France) (2012–2014), and an editor of Geophysical Research Letters (Solid Earth) (2001–2007). In 2007, the President of the Republic awarded him the honor of *Commendatore della Repubblica Italiana* for scientific merit. He is the author of more than 160 publications in high impact factor journals and of several internationally distributed books.

Ann Langley is Professor of Strategic Management at HEC Montréal, Canada, and holder of the Research Chair in Strategic Management in pluralistic settings. Her research focuses on strategic change, interprofessional collaboration and the practice of strategy in complex organisations. She is particularly interested in process-oriented research and methodology and has published a number of papers on that topic. In 2013, she was co-guest editor with Clive Smallman, Haridimos Tsoukas, and Andrew Van de Ven of a Special Research Forum of *Academy of Management Journal* on Process Studies of Change in Organisations and Management. She is also co-editor of the journal *Strategic Organisation,* and co-editor with Haridimos Tsoukas of a book series *Perspectives on Process Organisation Studies* published with Oxford University Press that is now on its 6th volume. She is Adjunct Professor at Université de Montréal, the Norwegian School of Economics and Business Administration (NHH), and University of Gothenburg.

Anthony C. Atkinson's undergraduate eduction was in chemical engineering at the University of Cambridge after which he worked for four years for Shell. During a spell at a manufacturing plant on Merseyside, he became interested in computing and statistics, especially George Box's work on experimental design for response surface models. A postgraduate year studying statistics at Imperial College London was followed by two years of working in the chemical industry in New Jersey. He then returned to Imperial College, where his PhD on model choice was supervised by David Cox. Thereafter, he remained at Imperial College, eventually becoming Professor of Statistics. In 1989, he moved to a similar position at the London School of Economics and Political Science, retiring in 2002.

His research interests are influenced by his early scientific training and industrial experience. A major activity has been in the design of experiments, particularly experiments that are applied in the chemical and other process industries. A second interest is in robust methods of data analysis, which are important when the data are badly corrupted or when they may be produced by a mixture of models. A third broad strand of work relates to the design of clinical trials in which the effect of randomization has to be explicitly included. The aim is to avoid biases and ensure that there is a balanced allocation of the various treatments to different kinds of patients.

Arthur M. Jaffe is the Landon T. Clay Professor of Mathematics and Theoretical Science at Harvard University. His research has spanned quantum field theory, statistical physics, probability theory, analysis, gauge theory, non-commutative geometry, supersymmetry, and reflection positivity. He has shown that quantum theory and special relativity are compatible in two-dimensional and in three-dimensional space–time. Recently, he has become interested in the development of mathematical picture languages and in quantum Fourier analysis. Applications of this work include a new approach to error correction in quantum computation.

In the 1970s, Jaffe was active in the foundation of the International Association of Mathematical Physics and served for six years as its President (1990–1996). He co-founded a sequence of summer schools on mathematical physics that met in Cargèse, Corsica, from 1976 to 1996. He acted as chief editor of *Communications in Mathematical Physics* for 22 years. Jaffe was Chair of the Harvard University Department of Mathematics, after which he became President of the American Mathematical Society. In 1998, Jaffe guided the conception of the Clay Mathematics Institute, then served as its President and designed all its programs, including the "Millennium Problems in Mathematics."

Jaffe received the Dannie Heineman Prize for Mathematical Physics from the American Physical Society and the American Institute of Physics, and the Prize for Mathematical and Physical Sciences from the New York Academy of Sciences. He is a fellow of the American Academy of Arts and Sciences, a member of the US National Academy of Sciences, and an honorary member of the Royal Irish Academy. He has also had a lifelong interest in classical music.

Bernard Sinclair-Desgagné is the International Economics and Governance Chair at HEC Montréal and is a faculty member at the École Polytechnique, Paris. Previously, he was for several years a Professor at INSEAD and the École Polytechnique of Montreal. He holds a PhD in Management Science/Operations Research from Yale University. His main research areas are the economics of incentives and organisation, environmental economics and policy, risk management, and environmental innovation. His publications can be found in major journals such as *Econometrica*, *Management Science*, and the *Journal of Environmental Economics and Management*. His recent work focuses on incentive compensation, global supply chains and the environment, the environmental goods and services industry, and policy-making under scientific uncertainty. In 2004, he was nominated a fellow of the European Economic Association. In 2006, he won (with co-author Pauline Barrieu of the LSE) the *Finance and Sustainability European Research Award* for the article "On Precautionary Policies" published in *Management Science*.

Sir Brian Hoskins is currently part-time Chair of the Grantham Institute at Imperial College London and Research Professor in Meteorology at the University of Reading. He is also a member of the UK Committee on Climate Change.

Until his retirement in April 2014, Sir Brian had been Professor of Meteorology at Reading for more than 30 years and was the Founding Director of the Grantham Institute. He held a Royal Society Research Professorship from 2000 to 2010. He was a review editor and an author in the Intergovernmental Panel for Climate Change Fourth Assessment Report. He served as Vice-Chair of the Joint Scientific Committee of the World Climate Research Programme from 2000 to 2004. He was President of the International Association of Meteorology and Atmospheric Sciences from 1991 to 1995. For many years, he was Chairman of the Met Office Scientific Advisory Group and was also its Non-Executive Director.

Professor Hoskins is a member of the Scientific Academies of the UK, USA and China. He has received the top awards of the UK and USA Meteorological Societies and is their honorary member. He has recently been awarded the Buys Ballot Medal and the inaugural Gold Medal of IUGG. He has honorary degrees from Bristol and UEA, and is an honorary fellow of Trinity Hall, Cambridge. In 2007, he was knighted for Services to Environmental Science.

Charles F. Manski has been Board of Trustees Professor in Economics at Northwestern University since 1997. He previously was a faculty member at the University of Wisconsin–Madison (1983–1998), the Hebrew University of Jerusalem (1979–1983), and Carnegie Mellon University (1973–1980). He received his BS and PhD in economics from MIT. in 1970 and 1973. Manski's research spans econometrics, judgment and decision, and analysis of public policy. He is author of various books, including *Public Policy in an Uncertain World* (Harvard, 2013), *Identification for Prediction and Decision* (Harvard, 2007), *Social Choice with Partial Knowledge of Treatment Response* (Princeton, 2005), *Partial Identification of Probability Distributions* (Springer, 2003), *Identification Problems in the Social Sciences* (Harvard, 1995), and *Analog Estimation Methods in Econometrics* (Chapman & Hall, 1988). He has served as director of the Institute for Research on Poverty (1988–1991) and as chair of the Board of Overseers of the Panel Study of Income Dynamics (1994–1998). Manski is an elected member of the National Academy of

Sciences, an elected fellow of the Econometric Society, the American Academy of Arts and Sciences, and the American Association for the Advancement of science, and an elected corresponding fellow of the British Academy.

Chris Impey is a University Distinguished Professor and Deputy Head of the Department of Astronomy at the University of Arizona. He has over 170 refereed publications on observational cosmology, galaxies, and quasars, and his research has been supported by $20 million in grants from NASA and the NSF. He has won 11 teaching awards, and he is currently teaching an online class with over 19,000 enrolled. Impey is a past Vice President of the American Astronomical Society and he has been an NSF Distinguished Teaching Scholar, the Carnegie Council's Arizona Professor of the Year, and a Howard Hughes Medical Institute Professor. He has written over 40 popular articles on cosmology and astrobiology, two introductory textbooks, a novel called *Shadow World* (2013, Dark Skies Press), and six popular science books: *The Living Cosmos* (2007, Random House), *How It Ends* (2010, Norton), *Talking About Life* (2010, Cambridge), *How It Began* (2012, Norton), *Dreams of Other Worlds* (2013, Princeton), *Humble Before the Void* (2014, Templeton) and *Beyond: Our Future in Space* (2015, Norton).

Gerd Folkers studied Pharmaceutical Sciences and attained a PhD in 1982 in pharmaceutical chemistry. In 1991, he was appointed Professor for Pharmaceutical Chemistry at the ETH in Zürich. The emphasis of his research was the molecular design of bioactive compounds and their application for a personalised therapy of tumours and diseases of the immune system. He served at the Research Council of the Swiss National Science Foundation from 2003 to 2011. From 2004 to 2012, he was Head of the Collegium Helveticum, a joint project of ETH Zürich and University of Zurich for the study of new scientific perspectives on transdisciplinary processes. Since 2012, Folkers has been a member of the Swiss Science and Innovation Council and has served as its Vice-President since 2014. In 2016, he was elected as the President of the Swiss Science and Innovation Council and appointed as Chair for Science Studies at the ETH Zürich.

About the Contributors xix

Leonard Allen Smith was raised in northeast Florida. He received his Bachelor's degree with honours in physics, mathematics and computer science from the University of Florida, and graduate degrees in physics from Columbia University in the City of New York where he worked under Ed Spiegel, one of the fathers of "chaos." His time as a GFD fellow at Woods Hole made a deep impact, both in raising his awareness of what "serious" science was, and how to deeply enjoy the process and the fellowship of science. In 1992, after postdocs at Cambridge and Warwick, he obtained a Senior Research Fellowship in Pembroke College, Oxford, which he still holds. In 2000, he joined Howell Tong's Centre for the Analysis of Time Series at the London School of Economics and Political Science (LSE), of which he is now the Director. He holds a tenured research Professorship in Statistics at the LSE, and is a Visiting Professor at the University of Chicago and at Durham University. Being a physicist employed as a statistician in the London School of Economics has allowed him to work with a rather wide range of nonlinear models, theories, computer programs, and somewhat related phenomena. His book *A Very Short Introduction to Chaos* is one of the bestselling technical volumes in the OUP series, and has been translated into over half a dozen languages. A sought after speaker, he is a Selby Fellow of the Australian Academy of Science. The Royal Meteorological Society awarded him their Fitzroy Prize for contributions to applied meteorology.

Michael Stumpf is a theoretician working on molecular and cellular processes in biology. Following undergraduate studies in mathematics, theoretical physics, and chemistry in Tübingen and Göttingen, he completed his DPhil at the University of Oxford. After this, he moved into theoretical biology, staying at Oxford, but in the Department of Zoology as a Wellcome Trust Research Fellow. After a year at the University College, London, he moved to Imperial College London in 2003, first as a Reader in Bioinformatics and then, in 2007, as Professor in Theoretical Systems Biology. Much of his work is concerned with the development of new statistical and computational approaches to develop better mechanistic, and predictive models, for example, gene regulation and signal transduction systems and the ways in which they shape the life (and development) of organisms. Here, balancing the complexity and usefulness (both in terms of capturing the important aspects and predictive ability) are challenging and much of his recent methodological work has been concerned with this problem. He lives with his wife and two children in West London.

Nigel Klein is Professor and Consultant in Paediatric Infectious Diseases and Immunology at Great Ormond Street Children's Hospital, London, and the Institute of Child Health, University College London. He trained at the University College London (UCL), obtaining degrees in anatomy and in medicine. He established and led the Infectious Diseases and Microbiology Unit at ICH until 2014 and then established and led the Department of Infection at UCL for five years until 2008. He has been working in the fields of Infectious Diseases for many years and has a particular interest in meningitis, sepsis, innate immunity, premature labour, and HIV. He has been a member of several boards, including the Wellcome Trust Infection and Immunity Board and the Chair of the Grants Panel for Meningitis, UK. Currently, he is a member of the Meningitis Research Foundation Grants Panel and Action Medical Research Panel. He has served on the Malawi Welcome Trust International Scientific Advisory Board and the MRC International Nutrition Group Scientific Advisory Board. He is a member of the PENTA Immunology and Virology committee. He is on the editorial boards of two journals and was a member of the Steering Group for the Technology Strategy Board's Innovation Platform on the Detection and Identification of Infectious Agents (DIIA). He has published more than 300 papers, many of them in high impact journals. And has also supervised more than 30 PhD students, of which 16 have been Clinical Fellows. Klein is currently on the Paediatric Medicines Expert Advisory Group of the Medicines and Healthcare Products Regulatory Agency and leads the Maternal, Child and Adolescent Programme in the recently funded Wellcome Trust Unified Institute in South Africa.

James Keirstead is Lecturer in the Department of Civil and Environmental Engineering, Imperial College London. He has a multi-disciplinary background, including degrees in civil engineering, environmental management, and energy policy and his current research focuses on urban energy systems. He has over 20 peer-reviewed publications in leading energy journals, has co-edited *Urban Energy Systems: An Integrated Approach* (Earthscan), and has contributed to major international assessments of urban energy use and its impacts. He is an elected Board Member of the Sustainable Urban Systems section of the International Society for Industrial Ecology and is a Chartered Engineer and member of the Energy Institute.

Nilay Shah is the Director of the Centre for Process Systems Engineering (CPSE). This is a multi-institution (Imperial and UCL) research centre which covers modelling and model-based design and optimisation of energy and process systems. It is one of the largest such centres in the world. Shah's research interests include the application of process modelling and mathematical/systems engineering techniques to analyse and optimise complex, spatially- and temporally-explicit, low-carbon energy systems, including hydrogen infrastructures, carbon capture and storage systems, urban energy systems, and bioenergy systems. He is also interested in devising process systems engineering methods for complex systems, such as large-scale supply chains and biorenewable processes, and in the application of model-based methods for plant safety assessment and risk analysis. He has published widely in these areas and is particularly interested in the transfer of technology from academia to industry. He has provided consultancy services on systems optimisation to a large number of process industry and energy companies.

Paul Embrechts is Professor Emeritus of Mathematics and co-founder of RiskLab at ETH Zürich, which specializes in actuarial mathematics and quantitative risk management. His previous academic positions include appointments at the University of Leuven, the University of Limburg, and Imperial College, London. Dr. Embrechts has held Visiting Professorships at the University of Strasbourg, ESSEC Paris, the Scuola Normale in Pisa (Cattedra Galileiana), the London School of Economics (Centennial Professor of Finance), the University of Vienna, Paris 1 (Panthéon–Sorbonne), the National University of Singapore, and the University of Hong Kong. He was Visiting Man Chair 2014 at the Oxford–Man Institute of Oxford University, and has honorary doctorates from the University of Waterloo, the Heriot–Watt University, Edinburgh, Université Catholique de Louvain and City, University of London. He is an elected fellow of the

Institute of Mathematical Statistics and the American Statistical Association, Actuary-SAA, honorary fellow of the Institute and the Faculty of Actuaries, Member Honoris Causa of the Belgian Institute of Actuaries and is on the editorial board of numerous scientific journals. He belongs to various national and international research and academic advisory committees. He co-authored the influential books *Modelling of Extremal Events for Insurance and Finance* and *Quantitative Risk Management: Concepts, Techniques, and Tools*. Dr. Embrechts consults on issues in quantitative risk management for financial institutions, insurance companies, and international regulatory authorities.

Ron Bates works for Rolls-Royce PLC as a Corporate Specialist in Robust Design and is responsible for managing uncertainty in the design and analysis of engineering systems. Part of this role involves working with the Rolls-Royce University Technology Centre (UTC) network of research universities to develop research and deliver new technology to the company. Prior to this, Bates helped set up the Decision Support and Risk Group at the London School of Economics and Political Science and was a senior member of the Risk Initiative and Statistical Consultancy Unit (RISCU), a research and consultancy group within the Department of Statistics at the University of Warwick. Bates is a specialist in developing and implementing methods for evaluating the robustness and reliability of multi-disciplinary engineering systems and processes, including contributions to the fields of design of experiments, spatial modelling methods (Gaussian process Kriging, Radial Basis Functions, Polynomial models), signal processing (including wavelet analysis), and multi-objective optimisation, all requiring a high level of computational modelling. He has published over 50 journal and conference papers, and presented papers and chaired sessions on Robust Design and Uncertainty Quantification at many national and international conferences. He is a Chartered Engineer and fellow of the Institution of Mechanical Engineers and a Chartered Statistician and member of the Royal Statistical Society.

Simon Dietz is one of the founders and currently the Co-Director of the Grantham Research Institute on Climate Change and the Environment at the London School of Economics and Political Science (LSE), where he is also the Director of the ESRC Centre for Climate Change Economics and Policy, and Associate Professor in the Department of Geography and Environment. Dietz is an environmental economist with diverse interests from climate change to biodiversity and from decision theory to growth theory. As an undergraduate, he studied environmental science at UEA Norwich and ETH Zürich before completing his MSc and PhD at LSE in environmental policy and economics. In 2006–2007, he was an analyst at the UK Treasury on the Stern Review on the Economics of Climate Change, and played a leading role in the Review's modelling of the "cost of inaction." He sits on the editorial boards of the *Journal of Environmental Economics and Management* and the *Journal of the Association of Environmental and Resource Economists*.

Stephan Hartmann is Professor of Philosophy of Science in the Faculty of Philosophy, Philosophy of Science, and the Study of Religion at LMU Munich, Alexander von Humboldt Professor, and Co-Director of the Munich Center for Mathematical Philosophy (MCMP). From 2007 to 2012, he worked at Tilburg University, the Netherlands, where he was Chair in Epistemology and Philosophy of Science and Director of the Tilburg Center for Logic and Philosophy of Science (TiLPS). Before moving to Tilburg, he was Professor of Philosophy in the Department of Philosophy, Logic, and Scientific Method at the London School of Economics and Political Science (LSE) and Director of LSE's Centre for Philosophy of Natural and Social Science. He is President of the European Philosophy of Science Association (EPSA, 2013–2017) and President of the European Society for Analytic Philosophy (ESAP, 2014–2017). His primary research and teaching areas are philosophy of science, philosophy of physics, formal epistemology, and social epistemology. He has published numerous articles and a book, *Bayesian Epistemology* (with Luc Bovens), in 2003 with Oxford University Press. His current research interests include formal social epistemology (especially models of deliberation, norm emergence, and pluralistic ignorance), the philosophy and psychology of reasoning, intertheoretic relations, and (imprecise) probabilities in quantum mechanics. He is currently working on the book *Bayesian Philosophy of Science* (with Jan Sprenger), which is under contract with Oxford University Press.

Contents

Preface vii

About the Editor xi

About the Contributors xiii

Chapter 1   A Conversation with Albert Ma   1

Chapter 2   A Conversation with Aldo Zollo   19

Chapter 3   A Conversation with Ann Langley   27

Chapter 4   A Conversation with Anthony C. Atkinson   37

Chapter 5   A Conversation with Arthur Jaffe   55

Chapter 6   A Conversation with Bernard Sinclair-Desgagné   71

Chapter 7   A Conversation with Sir Brian Hoskins   87

Chapter 8   A Conversation with Charles Manski   109

Chapter 9   A Conversation with Chris Impey   137

Chapter 10 A Conversation with Gerd Folkers   167

Chapter 11 A Conversation with Leonard Smith   199

Chapter 12 A Conversation with Michael Stumpf   233

Chapter 13 A Conversation with Nigel Klein   253

Chapter 14 A Conversation with James Keirstead   269

Chapter 15 A Conversation with Nilay Shah   285

Chapter 16 A Conversation with Paul Embrechts   297

Chapter 17 A Conversation with Ron Bates   311

Chapter 18 A Conversation with Simon Dietz   319

Chapter 19 A Conversation with Stephan Hartmann   333

Chapter 1
A Conversation with Albert Ma

Can you describe your field briefly?

I am an applied microeconomist, and my research is on health economics and industrial organisation. My main interest is in the health market, and my favourite way to describe my work is by means of a "three-circle" diagram. Each circle represents a "party": an insurer, a provider, and a consumer. The economics of the health market is about how these three parties interact.

Present a paradigmatic example of a model in your field, describing it in terms that are accessible to non-experts.

Obviously, the health market is complex; it accounts for about 20% of the GDP in the United States. There are many insurers, both public and private. There are many providers such as physicians, nurses, and allied health professionals. And there are about 330 million people in the U.S.! When we economists face such a complex situation, our angle of attack is to simplify. Indeed, every single piece of research in the health economics field is about some aspect of this three-circle diagram. However, there isn't much work that is about EVERY aspect of this diagram.

Economists are very good at simplifying things. Let me illustrate this with two examples. First, one line of research on the current health care reform in the U.S. focuses on the newly formed state and federal Health Insurance Exchanges, which are market places for consumers to buy health insurance. Of course, this simplifies the research but ignores the role of providers. Second, another line of research focuses on how physicians and hospitals organise themselves for better efficiency; this is encouraged by the health care reform. In this case, this simplification ignores consumers' insurance purchase decisions. We simplify in order to focus on a limited number of relevant issues so that ultimately a deeper analysis can be achieved.

With the help of these examples, could you explain why a model is needed? And could you describe what a model is, with regard to the different stages in the modelling process in particular?

I would argue that this is, sort of, the only game in town. When we set out to study a problem, we always begin with a specific focus. This requires us to get rid of issues that are irrelevant to our focus. Again, in the first example above, health exchange studies should not involve how pharmaceutical companies decide to advertise to consumers. And, in the second example, studies of the Accountable Care Organisations, the mode of health care delivery advocated by the U.S. health care reform, should not involve health insurance premium subsidies to those consumers with lower incomes.

What determines the boundary? How do we decide what to include and what to exclude in our research? That is when a "model" becomes relevant. I'd just say that a model is a collection of postulates, often accepted by most members of the research community, for carrying out analytical and empirical work. The collection of postulates is often quite small so that some issues are not mentioned at all. For example, the evaluation of consumers' uncertain incomes according to the expected utility hypothesis is often taken for granted. But postulates are the analytical foundations.

Modelling is the construction of a simplified account of the issue of interest. To study the Health Insurance Exchanges, we assume that consumers have different likelihoods of becoming ill, and each must purchase some insurance. More importantly, we assume that firms that sell insurance policies cannot tell them apart, so they must offer the same premium. Our model of an insurance market may begin with these two postulates. It has been well established that the market may perform poorly in this model. The role of the model is to explain how the postulates result in poor performance. Finally, we then use the model to study how Health Insurance Exchanges may reduce poor performance by regulating the insurance market.

Using the second example, we begin with the assumption that the delivery of health services may be complicated by insurance. Because consumers do not pay the full cost of care, they may not have the incentive to acquire knowledge about treatment effectiveness and costs, so often use excessive and expensive care. We also assume that, without the proper incentives, health care providers do not choose the most cost-effective treatments. Our model of a care organisation can simply be given by these two postulates. Now, the study of Accountable Care Organisations is concerned with how innovation in organisational structures can encourage cost-effectiveness.

The reason for a model is obvious. Without a model, we do not know where to start. Without a model, we do not know how to formulate a problem. Finally, without a model, we do not know how to evaluate propositions or recommendations. Statements or results follow from some postulates, so they are meaningful only if we have listed those postulates in the model.

In the first example, consider the policy recommendation of mandatory health insurance. According to the two postulates, mandatory insurance results in better risk pooling, so market failure due to missing information can be alleviated. However, if we had not made those assumptions, we would have to think of another reason, perhaps fairness, or equal treatment for all. One policy can be given multiple interpretations depending upon the model.

In the second example, consider the policy recommendation of integrated delivery of medical services in return for a fixed, capitated payment per enrolled consumer. According to the two postulates, capitation forces providers to be responsible for costs, and integration allows the alignment of incentives among different providers. Without those postulates, then, we would have to think of perhaps solidarity and social adhesion as the reason for the integration policy.

Economists subscribe to model building. But we have borrowed that from the scientists. The so-called scientific method is about models. Allow me to use perhaps one of the most amazing models in physics to illustrate. Einstein's special relativity consists of a single postulate: that the speed of light is constant to every observer no matter at which speed the observer is travelling. In Einstein's own (translated) words: "the same laws of electrodynamics and optics will be valid for all frames of reference for which the equations of mechanics hold good" (*On the Electrodynamics of Moving Bodies*, 1905). Einstein's model building can also be illustrated again by his work that paved the way for quantum mechanics. He said, "According to the assumption considered here, when a light ray starting from a point is propagated, the energy is not continuously distributed over an ever increasing volume, but it consists of a finite number of energy quanta, localised in space, which move without being divided and which can be absorbed or emitted only as a whole" (*On a Heuristic Point of View about the Creation and Propagation of Light*, 1905).

Whereas many regard model building as consisting of many structures and substructures, etc., I think of model building as something simpler than that. Indeed, typically there are various components for any model. Although Einstein's special relativity paper contained no reference, he did mention a few names: Newton, Maxwell, Hertz, and Lorentz. (The photoelectric paper did contain references, a total of four.) Each model is a picture of something that is of interest.

But in the models of relativity and the photoelectric effect, each one has a central canon of construction. And that, I think, is the key to understanding models. A model puts forward a central concept — the constant speed of light, and the discreteness of light and energy — in Einstein's models. Obviously, we know now that these were revolutionary concepts. (It can be said that Einstein knew them to be revolutionary before the rest of the world. See *Einstein: His Life and Universe* by Walter Isaacson, 2007. The daily lives and careers of most scientists have less drama.)

Similarly, the model describing Health Insurance Exchange has the postulates that consumers possess private information about the likelihood of themselves developing illnesses as its canon of construction. The model describing Accountable Care Organisations has misalignment of incentives as the canon of construction. These are probably less revolutionary than relativity and light quanta, but they have been central postulates for economists for the past 50 years, during which time information economics has taken the central stage in microeconomic theory.

Of course, before relativity and before light as quanta, there were other models about space and light (space and time being independent, and light being a wave). Likewise, before asymmetric information and incentive alignments, there were models about uncertainty, and models about incentive misalignment. For example, the general equilibrium model with uncertainty, and externality and public good problems are related theories. However, they have not had the asymmetric-information aspect so critically emphasised in information economics. The terms adverse selection, moral hazard, hidden information, and hidden action were not in the dictionary of the classical general equilibrium model.

What is the role of mathematics in modelling?

I think that mathematics has very little to do with the basic principles of a model. There is nothing very mathematical about consumers having private information about their likelihood of becoming ill. Likewise, it is not a mathematical statement to postulate that providers' incentives are misaligned. The first is about information that is known to some but not to others. The second is about how some economic actors will behave. They are hardly mathematical! They assert something that may or may not be true. Mathematical statements are seldom like that.

Even in physics, the statement that the speed of light is constant to an observer travelling at any speed is not a mathematical statement. Similarly, the statement that there is a smallest packet of energy is not a mathematical statement. These are statements about physical events or entities.

One can, in principle, find these statements to be false, and the validity of these statements has nothing to do with mathematics.

The key is that a model provides the concepts necessary for understanding some phenomena or relationship between phenomena. Thus, private information is the key to understanding why a market may fail, and incentive misalignment explains why decisions can lead to inefficient outcomes.

The constant speed of light dispenses with the assumption of a permeating substance through which everything else travels. The quantum theory of light dispenses with the formal difference between gaseous and electromagnetic processes.

However, that does not mean that mathematics is unimportant in modelling. A model begins with postulates, but we put in a lot more other structures to build up the model. The theory for the Health Insurance Exchanges is not just about consumers having superior information. The theory for the Accountable Care Organisations is not just about incentive misalignment. These are the most important postulates. Successful postulates should be deep enough to generate further developments. And this is the art of model building. One makes a seemingly simple postulate, but its implication can be enormous. Again, Einstein's postulates are probably the deepest in modern physics. The acknowledgement of asymmetric information and incentive misalignment is probably the deepest postulate in modern microeconomics.

That, I think, is where mathematics comes in. Mathematics is at its best when it is used to describe relationships. I think that the dominance of game theory in economics is precisely because of that. Game theory is a framework that is extremely good for describing interactions between individuals. The strategic aspects in any relationship can be made precise by game theory. Mathematics allows us to use the depth of postulates in a model. One can say that mathematics is a tool, but it is the most valuable of all tools. How would modern economics (or physics for that matter) function without mathematics? It would be unimaginable.

Hence, in the model for Health Insurance Exchanges, we need to describe how consumers and insurers interact. More importantly, we need to describe the strategic role of asymmetric information. When insurers compete, say by offering policies at various premiums, they must recognise that consumers will make sure of their superior information. Our model therefore must describe precisely how consumers can exploit their advantage. Mathematics does allow us to specify precisely what an insurer can offer and at what time; it also allows us to specify how a consumer can react based on his information. For example, the model can simply say that each insurer offers an insurance policy that consists of a premium and a level of coverage of medical expenses, and then, each consumer picks a policy from the available ones.

In the model for Accountable Care Organisations, we need to describe the precise degree of incentive misalignment. Suppose that there are general practitioners and specialists, and they have to decide how a patient is to be treated. Clearly, providing treatment requires resources and effort, but the determination of the best provider requires diagnosis. Will a general practitioner have the incentive to diagnose and refer a patient to a specialist purely based on cost-effectiveness? Likewise, will a specialist diagnose and refer a patient to a general practitioner for cost-effectiveness? The complete model will have to specify the financial incentives and professional concerns among physicians. It must also specify the consequences of referrals, both to physicians and patients. A game-theoretic model again specifies which physician can do what and at what time. It also specifies possible reactions by other physicians and possible outcomes for patients.

A mathematical model spells out the precise strategic interaction in any relationship. I can liken it to a play. There are actors (providers, insurers, consumers), and then there are the sets (the market, regulated or competitive), and then there are rules that the actors must follow when the play unfolds (what kind of contracts are allowed in the market, what are feasible regulations). Of course, in a play, actors may not exactly follow the director's instructions. In a scientific model, all actors must follow rules that are well specified and accurate.

Precision facilitates discussions, revisions, and new developments. It's just a matter of ordinary academic life and scholarly pursuit that researchers argue about this and that. Well, it's better that we can all agree on what we are arguing about — even though we may not agree!

There is also a practical advantage of mathematical models. Mathematical theorems are proven propositions. They are universal; their truth or falsehood is context-free. If my model employs a certain optimisation problem, and some theorem tells me the characterisation of the solution, then that's it. I have found the solution — or others have found it for me. No ifs, ands or buts! Mathematical theorems are never wrong. They can tell me what my results are — even if I don't understand them! And that's the big question. Sometimes we may know the solution, but we may not be able to comprehend what it means. I like to say to my students, "Mathematics is never wrong, but it won't tell you anything. It's your responsibility to understand the mathematics."

What constituents besides a mathematical formalism are part of a model?

As I said, the foundations of a model are almost always not mathematical. How these principles are to be used requires other building blocks. And often, the scaffolding to put everything together requires mathematics. It is as if mathematics provides all the linkages.

Allow me to use a construction example. We have a foundation of a building. Then we need to decide how tall it can be, and whether an elevator or an escalator can be used for transporting occupants up and down. Then we may have to decide on partitions, electricity, plumbing, windows, etc. Obviously, all these require compatibility. A two-story building may well have escalators as the only means for travelling up and down, but an elevator is better when the building is 10 stories tall. Then if we build a warehouse, the partitions and amenities aren't that critical. But if we build an office, then windows and electricity are more important, but elaborate showers and kitchens are not needed. However, if we build an apartment complex, then partitions and kitchen designs are very important.

So if we have a model that says consumers have superior information about their health conditions, we still have to decide on the other components of the model. Are there going to be many firms, or just one firm? The former will be about a market, so we should describe how firms interact, and what kind of insurance policies are allowed. The latter setup will be about a monopoly. Often in market situations, we let insurance policies be quite simple: one combination of premium and coverage is to be offered by each firm. But for a monopoly, we may want to allow more complex policies, such as a menu of premium-coverage pairs. Finally, if we want to investigate the long-term effects of asymmetric information, then we consider many rounds of interactions between consumers and firms.

Again, if we have a model that says providers have incentive misalignments, we still have to decide how they interact. Are there going to be many general practitioners interacting with many specialists? What kind of instruments, or financial arrangements, can providers use to solve incentive misalignment problems? Can they hire an auditor? Can they keep track of referrals and their outcomes, either clinical, financial, or both?

The constituents are not quite mathematical; they are the components necessary for a model to focus on one, or a few, issues. The model begins with some basic principles, and then we build it up. There are many, many ways to go. However, there is a limit as to how components can be fitted together. One is not going to construct a building that can withstand saltwater corrosion; that requirement — corrosion resistance — is for boats, not buildings. However, the components have to be fitted together in a mathematical way, just like various floors of a building must obey the laws of physics in terms of balance, weight, height, etc. Likewise, insurance policies offered by firms have to be defined mathematically, and they must also respect the missing information constraint (which usually means a certain measure-theoretic restriction).

What is the role of language in modelling?

I guess that by this time, I have given the impression that economic modelling is not that mathematical. Many models are described verbally. Of course, there are mathematical notation, symbols, expressions, formulas, equations, etc., but the economist just links them together using a common language.

More importantly, the professional vocabulary — the jargon — is English, so the meaning and connotation typically have everyday origins — in the English-speaking world. A native English speaker has an advantage when it comes to writing economics papers — everything else being equal.

We use a language to communicate our thinking. In the current academic environment, English happens to be that universal language for economics papers. Economic modelling requires critical and clear thinking. Hence, as long as a language can be an efficient vehicle to communicate critical thinking, it would be enough. No matter which language we happen to use, if thinking is muddled, then only muddled ideas can be communicated. Communicating muddled ideas clearly isn't exactly interesting, is it?

There is a more subtle issue, however, about the role of language. This concerns consistency. One can express mathematical relationships symbolically. If one has a suitable dictionary for the mathematical notation, then one can represent assumptions and results entirely in terms of mathematical notation alone. Now, restating those assumptions and results verbally typically helps in understanding. Sometimes, it is not just a restatement. Sometimes, it is an interpretation given in everyday language. The role of language is therefore to make that interpretation possible.

I want to liken this to some scenes in the 2003 movie *Lost in Translation* by Sofia Coppola. There were quite a few scenes in that movie in which the Japanese dialogue was very long, say over half of a minute, but the translator for Bill Murray would just say something in English for about 10 seconds. Well, is that good or bad translation? It is hard to tell — without knowing the context. But if we knew the exact context, why would we need translation? Very often, a result in an economics paper requires a very lengthy proof, but the discussion of the results can be quite succinct, even short. The author has glossed over many details. Now is that good or bad? Again it is hard to tell. If we could tell, we would have understood both the mathematics and the verbal description; often we don't have the time, or don't want to put in the effort to do so.

So would you say that language plays a role in modelling and also at the time of the communication of the findings? How important is notation?

This is a bit of a strange question for me. Although I speak Cantonese and English, I have never thought about economics in Cantonese. I have never spoken about economics in Cantonese either. It would be impossible for me to present an economics seminar in Cantonese. So the question is almost like asking what it is like navigating a terrain like a bat. I have no experience of using a language other than English in economics!

Notation is the ultimate deal breaker. As Shakespeare said, "Brevity is the soul of wit," I'd say, "Notation is the soul of comprehension." The usual human short-term memory limitation is about seven items. If I see an equation consisting of five Greek and three Latin alphabets on the left-hand side, and another five or six similar entities on the right-hand side, I'd be lost. I just can't comprehend so many things. Such a cumbersome misfortune, however, originates from an ill-conceived model. Why would one need about sixteen symbols to have some ideas pinned down? That's usually an indication that there has been some confusion. Clear thinking usually leads to simple notation.

Models are often said to represent a target system (typically a selected part or aspect of the world). Does this characterisation describe what happens in your field? If so, could you say how a model represents its target? In other words, how do you understand the model–world interface?

Models are simplifications. What it really means is that any model must maintain a certain focus — the target. This is true in almost any scientific discourse, and of course in my own research. Recall the three circles that I put up earlier. Sometimes I'd even focus on one single circle, say the provider. Then I will have to ignore or heavily simplify the description of the insurer and the consumer. For example, I can assume that the insurer sets a price for a medical service for each provider in the market, and that consumers' behaviours are described by a demand function. Then I can focus on how a provider reacts to the price and the demand. For example, I studied provider quality and cost-reduction efforts in a paper of mine in 1994. I even extended it to a provider dumping costly patients and cream-skimming profitable patients. The fixed-price assumption for the insurer and the abstract demand function were very convenient; the analysis did not get too complicated, but the main principles were readily understood.

However, the simplification does ignore some elements that are of interest in a broader context. Obviously, more complex interactions between providers and consumers cannot be accommodated. More elaborate insurer pricing rules, such as risk adjustment and pay for performance, have been set aside.

But this is the way things work in any scientific endeavour. There was first special relativity, and then there was general relativity, more than 10 years later. Motion without acceleration was easier to study, but that study was sufficient to change, once and for all, our conception of time and space. Motion with acceleration was more difficult to study, but when it was done, our notion of mass and space was completely changed. I am not familiar with all histories of science, but it seems that scientific models typically begin as simple ones, and then gradually become more complex.

What is the relationship between a model and a theory?

A theory means a set of related models. As I just said, we have special relativity and general relativity. Together, they formulate a theory about the macroscopic universe, as we know it.

My three-circle diagram can be thought of as a set of models. I have already mentioned the circle that represents the provider. Next, we can consider the insurer, and with some simplification of the roles of the provider and the consumer, we have a model about adverse selection in the insurance market. We can even explore advantageous selection, the opposite of adverse selection, in a modified setup.

Finally, we can consider the consumer after we simplify the setup for the provider and the insurer. We then can study how consumers respond to information provided by pharmaceutical companies, or how patients with chronic conditions respond to choices between general and speciality care.

Now, we can consider the two circles together. Giving consumers a simplified choice set by postulating that most consumers trust doctors to decide on their treatments, we are going to study how an insurer can influence a physician's decisions. Can a typical prospective payment contract motivate a physician to choose the right care quality? Or is it a more elaborate partial fee-for-service schedule supplemented by capitation? How does each compare to cost reimbursement? What are the effects on cost efficiency? Alternatively, we can abstract from the insurer and study interactions between physicians and patients. We can study induced demand, compliance, and collusion.

The point is that the basic three-circle diagram embeds in it many model variations. Depending on the focus, some interactions are highlighted, whereas some are suppressed. The grand model, the one that depicts of how the three parties interact with each other, is also available. Admittedly, such a model tends to be more complex, but it is there. All these models form a theory — an economic theory — of the health market. And I emphasise that this is an economic theory. I can imagine that one can set up these interactions in a framework of medical ethics, sociology, and even political relationships between participants in the health sector. But economists have less convincing expertise about these other aspects!

What is the aim/use of the model: e.g. learning/exploration, optimisation/exploitation?

One can be scholarly about the aim of a model or a theory. G. H. Hardy, a British mathematician who, together with J. E. Littlewood, revolutionised modern mathematics, was so proud of his mathematics being utterly useless — in the same sense that a chess game is useless. But economics isn't mathematics, and most economists will find it odd — if not entirely depressing — if what they do is found to be as useless as a chess game. (I must say that chess games have proven to be very instrumental in our understanding of artificial intelligence, bounded rationality, etc., so they are far from being useless. And of course, Hardy was also wrong about his mathematics being utterly useless.) But in a scholarly pursuit, one is not to look at immediate usefulness as a goal of developing a theory. In fact, an insistence on usefulness might just turn out to be short-sighted.

Perhaps the most complete set of economic theory is the general equilibrium model, which seeks to find conditions for the existence of a competitive equilibrium, and its efficiency property. Kenneth Arrow, Gerard Debreu and many others did this together in the 1940s and 1950s. There was an elegance to it; it was a most beautiful model of the market. What was the aim of the general equilibrium model? Finding a set of conditions for the harmonious operation of the free market, or ascertaining the efficiency properties of its equilibria? I am not sure if researchers at the time could have universally agreed on an aim. But scientists and scholars work best when they are free to think and create, so a uniform aim for a model is probably counter-productive. But they did it. My point is that an agreed aim for a model may not be needed. The market of ideas perhaps can sort itself out.

Now a lot of economists, and especially health economists, are interested in developing models to inform and enlighten policies. That's when things become tricky. Models are supposed to be simplified versions of reality. Policy is about the real world, which has so many confounding elements. Conclusions and recommendations from models may fail to be useful precisely because those models have ignored some significant parts of reality.

If the aim of a model is to inform policy, the model must be relevant. To study health care reform, a model of the insurance market must include the regulations in Obama Care. The researcher has no other alternative. The model, therefore, tends to be less general. It would not apply to another economy, such as the one in Canada, or Sweden. My point is that one has to be practical if the aim is a practical recommendation!

When you use computer simulations, what is the relationship between the simulations and the model?

Well, I don't do simulations, but can appreciate their power! Simulations are examples, which may eventually lead to insights, or give clues that can be used to formulate general results. But I have very limited experiences, so would not want to venture too far.

What is a good model?

This is a very deep issue. I tend to use a rule that is popularly attributed to Einstein: "A model should be as simple as possible, but not simpler." The good thing about a simple model is that it has just enough components to promote a good understanding. The general principle is that if you can use a model that employs two components to explain why fee-for-service pricing tends to raise health care costs, that's better than a model that employs twelve components! If you can make the point with one provider, why bother with a model with two providers?

However, economic models are not just about being simple and elegant. Economics is also about relevance. And that may require us to build models with complex and tedious components. Our model may have to incorporate key components of the real world. Sometimes the way to do it is to use an overly simplistic model, derive the results, and then argue that the confounding factors are not causing serious damage. If you want to inform health care policy, then some complexity must be expected. An overly simplistic model has the serious liability of neglecting too many things.

I think that what qualifies as a good model is a practical matter. Our work has different purposes, so each research study may require a different kind of model. It's too preposterous to aim for a theory of everything. The economics profession is mature enough that we can attempt to theorize on many social issues.

What has been the impact of the development of new technologies or tools (e.g. telescopes in cosmology, or computer simulations for DNA)?

That's too speculative for me to contemplate...really that's beyond my comfort level!

How would you define risk? And how does the model help us understand risk? Also, how would you define uncertainty? And how does the model help us understand uncertainty?

I am going to combine these two issues together. First, there have been a lot of studies on risk and uncertainty. Decision theorists spend their professional lives tackling the modelling of risk and uncertainty. Game theorists spend their professional lives tackling games of imperfect and incomplete information. Health economists deal with insurance markets in which the likelihood of a consumer getting sick is unknown. All the information economics literature from the past 50 years has been based on risk and uncertainty. There really is no need for me to add to this discussion.

However, I want to use this space to talk about uncertainty in economics. Any scientific discipline has an overarching framework. Particle physicists use their "standard model" to identify fundamental forces and particles. And relativity and quantum mechanics, although not entirely compatible, are widely accepted.

Now if we go back in time to before 1900, then we didn't have all these fancy theories in physics. The Newtonian theory of gravity was the only framework. And if we go backward for another 25 years before that, the relationship between electricity and magnetism had just been enunciated. Hence, before 1875, nobody would venture to say that electricity and magnetism were one and the same thing.

The point is that scientific advances cannot be easily predicted. Once in a while something significant will happen. In the past 100 years, perhaps the most significant single development in economics happened in 1936, when John Maynard Keynes published *The General Theory of Employment, Interest, and Money*. It was a theory that could not be cast in the neoclassical framework of Marshall, Pigou, and Walras. Indeed, Keynes wrote in the preface of *The General Theory* that his book was meant to be a message to his colleagues at the time. The neoclassical theory of demand and supply with price equilibrating the market, he said, would fail miserably. At that time, the notion that price could not equilibrate the market was so strange. Of course, now we would not be able to imagine what economics would be like if macroeconomics was deleted from the discipline!

Similarly, before the advance of game theory, the standard model of interaction was the competitive or imperfectly competitive market. The notion of an equilibrium used is specific to each context. The development of game theory introduced an entirely new language and set of tools for analysis. We now have a unified framework to study interactions between two people, two firms, many people, many firms; single or repeated interactions; with or without uncertainty, etc. Now we can't imagine what microeconomics would be like without game theory!

Allow me to paraphrase, "There are the known knowns, the known unknowns, and then the unknown unknowns!" (often attributed to a former US Defense Secretary, although the idea of unknown unknowns has been known to be used in NASA). What would macroeconomic models look like if we were not to use the current ones? What would microeconomics models look like if we were not to use game theory? The current models of the health sector use basic concepts such as selection, risk adjustment, and moral hazard. It is just impossible to imagine what health economics would be like if we were not to use those concepts.

But if we have learned from our disciplines, we can be sure that there are always new models and new theories. To me, that's the real uncertainty. When will new models come about? What will they be like?

What do you consider to be the work/result that has had the most significant impact on your field? And why?

This is a very difficult question. Health Economics is a big field. Ken Arrow's paper in the American Economic Review in 1963, "Uncertainty and the Welfare Economics of Medical Care" has often been said to be the first piece in modern health economics. That paper discussed a lot of issues, proposed some solutions, and presented a lot of insights. But what came after in the half century was a huge amount of research. Theoretical research, empirical research, experimental research, policy evaluation, and service discussions are all in the broad portfolio of health economics research. The diversity of research is a very reassuring sign. It means that the field has attained some maturity!

I think that the Handbook of Health Economics, all three big volumes of it, together with other editions of surveys published to date, show the significance of the field and its impact. Scientific research is very much serious team work! I would be remiss to name only a paper or two.

Chapter 2
A Conversation with Aldo Zollo

Can you describe your field briefly? Present a paradigmatic example of a model in your field, describing it in terms that are accessible to non-experts.

The basic model example I want to propose is the theory of seismic wave propagation which represents the Earth's internal structure in terms of spatially variable elastic properties such as rigidity or density. Actually, it is better if seismologists use seismic velocities (e.g. the speed of primary P and S waves in Earth rock materials) and density as the parameters to describe the elastic behaviour of geological media. Depending on the observed parameter to reproduce, a model can show the generally nonlinear, mathematical relationship between data and model parameters. The travel times of seismic waves propagating from a deep earthquake source to the receivers deployed at the Earth's surface are typically the physical quantities which seismologists measure and use to reconstruct elastic images of the Earth's interior in terms of spatially varying seismic velocities. In this case, the model is the theory allowing us to compute the travel time of the seismic waves. This requires the estimation of the wave trajectory and knowledge about the local variation of the seismic velocity along the travel path. In a simple uniform velocity medium, wave trajectories are straight lines, so the model is a simple linear equation ($T = \frac{r}{v}$, T is the travel time, r is the travelled distance and v is the seismic wave velocity). In a more complex geological medium, the seismic velocity can vary according to the different rock materials encompassed by seismic waves. In this case, the models become an integral equation ($T = \int_{path} \frac{ds}{v(x, y, z)}$, where ds is a small portion of the ray trajectory, $v(x, y, z)$ is the velocity field) where the integral extends to the wave trajectory and the total travel time is obtained from the integral of partial travel times along a small portion of the wave path.

With the help of this example, could you explain why a model is needed? And could you describe what a model is, with regard to the different stages in the modelling process in particular? In doing so, please answer the following sub-questions:

(a) **What is the role of mathematics in modelling?**
In 3D seismic imaging problems, the model is expressed by mathematical equations, which relate the parameters to be determined (e.g. the seismic velocity values at the nodes of a discretised geological medium) to the data (e.g. the arrival times of seismic waves generated by earthquakes or artificial explosive shots). The equations relative to all measured source-to-receiver travel-time data form a huge system of equations which is solved with the help of numerical, matrix inversion techniques implemented in small, medium, or large computer facilities, depending on the size of the geological medium and the expected image. So, the role of mathematics is relevant not only in model development, but also in its numerical implementation in computer algorithms which allows us nowadays to reproduce and simulate the real world with extreme accuracy.

(b) **What constituents besides a mathematical formalism are part of a model?**
The model parameter space of the seismic imaging problem includes not only the unknown parameters to be determined along with their uncertainties but also a number of known parameters which describe, for instance, the geometry of the data acquisition experiment. In the case of seismic tomography, which is a specific seismic imaging method, the location of receivers and sources, the size and

number of cells in which the medium is discretised, the topography of the Earth's surface where the receivers are deployed, the reference seismic velocity model, etc. are *a priori* known parameters of the model which are the needed ingredients of the model. The representation of the Earth's structure requires the set-up of the physical dimensions of the geological medium where the seismic wave propagation process is simulated. This requires the building of a discretised, synthetic geological medium reproducing the real-world variability of elastic properties, and the definition of the size of the elementary 3D cell or pixel which will compose the final subsoil Earth image.

(c) **What is the role of language in modelling? Are there qualitative aspects in modelling?**

The qualitative aspects that I see in the example of seismic wave propagation rely on the transposition of the geological idea of the subsurface Earth structure into the numerical models which are built to reproduce the data or to simulate the wave propagation processes. There are limits to the resolution of the data on retrieving reliable models, and therefore to our ability to simulate the details of Earth's structure using said data. In these cases, extrapolation/interpolation of reality is accomplished using qualitative or "trial and error" modelling approaches, which start from simple geological model representations and move toward more and more complex ones. The combination of different *a priori* datasets including speculative ideas about the model to be reproduced are essential in this preliminary reference model search.

(c′) **How important is notation?**

Notation is a part of the mathematical language that allows a better/optimised model description and compact formalism. In the case of complex, large-scale physical models of the Earth, notation is important but not essential.

(d) **Models are often said to represent a target system (typically a selected part or aspect of the world). Does this characterisation describe what happens in your field? If so, could you say how a model represents its target? In other words, how do you understand the model–world interface?**

Models are generally a simplified representation of reality, and generally a number of assumptions are to be verified for the model being appropriate to reproduce the real world. These assumptions normally originate from data resolution, in other words the limitations of instruments, sensors or methods to catch the details of the real processes. This is exactly the case for the seismic tomography method, whose capability to retrieve details of the Earth interior' geological structure critically depends on some data characteristics, such as the acquisition layout, the frequency content of seismic signals, the allowable medium discretization. Matching the new observations with simulated data, e.g. data numerically generated using the retrieved or reconstructed models, is the common practice used by geo-scientists to understand the extent at which their models are capable to reproduce the real world. A spectacular example is given by the observed travel times of seismic waves propagating within the Earth structure, which are well described by the one-dimensional Preliminary reference Earth model, with few percent deviations which are associated with the lateral earth heterogeneity.

(e) **What is the relation between a model and a theory?**

In Geophysics, the word "model" is used to express two concepts. "Model" is the theory relating data to parameters that scientists use to reproduce (or simulate) the observations or to determine the parameters of the theoretical relations. "Model" is also used for the three-dimensional description of the geological medium, e.g. the spatially variable rock physical properties. In classical seismic imaging problems, a model/theory of wave propagation is used to reconstruct a model of the subsurface Earth structure in terms of space-varying seismic velocities.

(f) **What is the aim/use of the model: e.g. learning/exploration, optimisation/exploitation?**

The model/theory is generally used to solve the "forward problem" in a typical optimisation problem, aimed at determining the best-fitting model parameter values. It is used to compute the theoretical (or synthetic) data that are compared (or matched) to real data in order to validate or reject a given set of model parameters.

In seismic imaging, the "forward model" solution is the core-component of a more general "inverse problem" aimed at finding the optimal space distribution of medium elastic properties, in other words, the image of Earth's interior in terms of elastic property contrasts.

The inferred representation of the discretised Earth is denoted as a "model" (Earth Model), which can be 1D, 2D, 3D or even 4D if time changes of rock material properties as represented by time-evolving seismic images are analysed in the time-lapse mode. The elastic Earth Model is used, for instance, to simulate wave propagation and also other physical processes such as earthquake ruptures, fluid storage and migration, land instability and deformation, etc.

(g) **In case you use computer simulations, what is the relationship between simulations and the model?**

In seismic imaging, the theory of wave propagation is translated into computer algorithms which allow us to evaluate the space–time evolution of a perturbation in a physical quantity, such as the displacement from a source to a number of receivers embedded in a 3D discretised volume. Each node of the volume is assigned a specific value of the elastic properties as represented by the primary (P) and secondary (S) seismic wave speed and density. The distribution of elastic properties in the 3D discretised medium is the Earth model. The theory/model is the mathematical differential equation governing the wave propagation whose local solution (obtained by numerical methods) allows us to simulate the wave propagation with the discretised 3D medium.

(h) **What is a good model? And, What has been the impact of the development of new technologies or tools (e.g. telescopes in cosmology, computer simulations for DNA, etc.)?**

Assessing "what a good model is" is really the most important aspect of any "inverse" (or optimisation) problem in Geophysics and Seismology. In the case of seismic imaging methods, like tomography, the quality of the retrieved model is basically referred to the goodness of fit. Given a norm (or cost) function which measures the distance between real and simulated data, the quality of a model relative to many others is assessed by the comparison of their "cost." The model which minimises the cost function is generally selected as the best one. But assessing the model resolution and uncertainty is also a big issue in model solving problems. In seismic imaging problems, resolution (which corresponds to evaluating the minimum pixel size that data are able to resolve) is investigated through specific tests using synthetic data generated in a slightly perturbed final model.

How would you define uncertainty? And, how does the model help us understand uncertainty?

The uncertainty on model parameters in a seismic imaging problem is related to errors in data (e.g. errors in the measurement of seismic wave arrival times) and inaccurate sampling of the acquisition layout (e.g. the sparse and poor distribution of source and receivers in the volume to be imaged). In seismic tomography, evaluating the uncertainty on model parameters, i.e. the error of the retrieved seismic speed at the grid nodes, is a difficult task and requires massive computations. A variety of methods are used (as the jackknife of bootstrap methods) based on mapping the data errors in the model parameter errors by repeating the seismic imaging process many times using different noised data series.

How would you define risk? And, how does the model help us understand risk?

Not clear about the concept of risk here.

For you, what is the experience/result in your field their has had the most significant impact? And why?

The most exciting seismic imaging problem that I have approached in my research activity is the reconstruction, detection, and imaging of the deep structure and magma feeding systems of the volcanoes Mt. Vesuvius and Campi Flegrei in southern Italy.

Starting in 1994, I was in charge of leading and coordinating a series of active seismic experiments on Mt.Vesuvius (TOMOVES) and Campi Flegrei (SERAPIS) volcanoes, in the framework of research activities aimed at mitigating the volcanic risk in the most densely populated and hazardous volcanic regions of Italy. This decade-long program of seismic investigation of Neapolitan volcanoes was solicited and promoted by the Italian Ministry of Civil Protection with the aim to integrate volcanic hazard studies based on the detailed imaging of the volcano structure and related eruption scenarios with data-based models of urban infrastructure (e.g. transportation, energy, telecommunications, water distribution, and public health), and with the social and political framework.

The high resolution Italian volcanoes imaging program that I and my scientific partners carried out during the last decade resulted in a major advance in the reconstruction of a complex structure, such as the interior of an active volcano, and had a wide impact on the worldwide volcanological community. The seismic evidence for a nearly flat, mid-crustal, primary magma reservoir feeding the eruptive activity of Mt. Vesuvius and Campi Flegrei has challenged the previous idea about a very shallow, nearly spherical magma chamber feeding the eruption activity and providing new insight into the deep plumbing system of volcanoes and their upward migration mechanisms.

Chapter 3
A Conversation with Ann Langley

Can you describe your field briefly?

I do qualitative research on strategic management processes and practices in complex pluralistic organisations, focusing on phenomena such as organisational change, the role of management tools, decision-making, leadership processes, and organisational identity, drawing on organisation and management theory.

Organisation and Management Theory (OMT) more generally is defined in the domain statement of the OMT Division of the Academy of Management as follows: *Organisation and management theory involves building and testing theory about organisations, their members and their management, organisation–environment relations, and organising processes.* http://aom.org/Divisions-and-Interest-Groups/Academy-of-Management-Division---Interest-Group-Domain-Statements.aspx#omt.

Present a paradigmatic example of a model in your field, describing it in terms that are accessible to non-experts.

The models I am interested in are conceptual representations of organisational processes. They are usually presented in a diagrammatic form and focus on various kinds of relationships among concepts. A classic

28 *Dialogues Around Models and Uncertainty*

model for understanding organisational change comes from a paper by Greenwood and Hinings[1] (Figure 1).

This model describes how changes in the market and institutional context of the firm or organisation may penetrate specific organisations in a given sector through two kinds of dynamics called "precipitating dynamics" and "enabling dynamics." Specifically, the authors argue that organisational change is precipitated when the interests of organisation members are not met, and when leaders have competitive or reformative value commitments that suggest a need to change, and that may in themselves be influenced by the context. While organisational change may be precipitated by such dynamics, the authors argue that change initiatives will only be successful if corresponding "enabling dynamics" are present. These include a distribution of power ("power dependencies" in Figure 1) that will enable reformist leaders to pursue their goals, and a configuration of organising capabilities and competences ("capacity for action" in Figure 1) that ensure that organisation members know how to mobilise resources to achieve change. The model suggests that when multiple organisations in a sector implement similar kinds of radical changes, this can recursively lead to transformation of the market and institutional context. Note that in this model, the market context refers to economic, technological, and competitive features of the industry in which the firm is located. The institutional context refers to the legal, normative, and taken-for-granted rules concerning how firms ought to behave to be considered legitimate in their field.

With the help of this example, could you explain why a model is needed? And could you describe what a model is, with regard to the different stages in the modelling process in particular?

This model is helpful in understanding when, why, and how organisations tend to radically change their structures, organising principles, and interpretive schemes in a given sector of the economy or institutional field. The model identifies key concepts in the process (items in the boxes) as

---

[1] Greenwood, R. and Hinings, C. R. 1996. Understanding radical organisational change: Bringing together the old and the new institutionalism. *Academy of Management Review*, 21(4): 1022–1054.

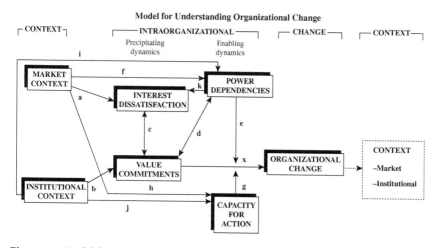

Figure 1. Model for Understanding Organisational Change.
Source: Greenwood and Hinings (1996).

well as the relationships between these concepts (expressed by the arrows in the diagram). The arrows may imply a "causal" relationship, but other kinds of relationships, such as "precedence" or "equivalence" or "mutual interaction," may also be expressed using different types of arrows (Langley & Ravasi, 2019).[2] There are no standard conventions for this, but the nature of the relationships as well as the concepts would be explained in detail in the text or in a legend. In the particular model shown in the figure, each relationship is associated with a specific formal proposition in the text. The usefulness of the model lies in its capacity to describe processes and relations among concepts that can be applied to a variety of generic contexts.

A given conceptual model is, however, only one possible representation of the processes involved. For example, there exist a variety of other models of organisational change in the management and organisation theory literature. The difference between specific models may be related to different levels of analysis (individual, group, organisation, population of organisations), and/or the mobilisation of different kinds of concepts to explain a phenomenon. Based on a detailed review, Andrew Van de Ven

---

[2]Langley, A. and Ravasi, D. 2019. Visual artifacts as tools for analysis and theorizing, In T. Zilber, J. M. Amis, and J. Mair (Eds.) *Research in the Sociology of Organizations*, 55: 173–199, Bingley, UK: Emerald Publishing.

and Scott Poole[3] indeed argued that models of change and development can be classified into four different canonical forms expressing the operation of different fundamental drivers: life cycle models (that consider patterns of change to be genetically embedded), teleological models (that emphasise the role of human agency and learning), evolutionary models (inspired by Darwinian processes involving elements of variation, selection and retention), and dialectic models (emphasising the role of conflict and contradiction in driving change). They suggest that most specific models in the organisational literature can be understood as involving combinations of these four canonical processes. While Van de Ven and Poole's typology of models is most useful, I would not be so certain that their four canonical processes cover the entire domain of the possible. However, one could argue, for example, that the processes represented in Figure 1 include elements of evolutionary processes reflected in the impact of environmental forces, elements of teleology reflected in the notion of interests and value dissatisfaction, and elements of dialectical processes reflected in the emphasis on power dependencies. This model does not however incorporate life cycle elements. In contrast, there exist several life cycle models of organisations (e.g. incorporating chronological stages of, for example, emergence, growth, maturity, shake-out, and decline).

What is the role of mathematics in modelling?

In my field, mathematical representations (e.g. through linear equations) might be used to test models such as the one given in Figure 1 using regression techniques or structural equations. However, such testing would require a means to "measure" or "assess" every item in the model, and to confront the relationships expressed in it with empirical data. Given the temporal and conditional relationships in this particular model, and the categorical rather than interval scale nature of the concepts, a mathematical formulation of the model above might not be most appropriate.

---

[3] Van de Ven, A. H. and Poole, M. S. 1995. Explaining development and change in organisations. *Academy of Management Review*, 20(3): 510–540.

What constituents besides a mathematical formalism are part of a model?

The model shown in Figure 1 was actually developed inductively from a major qualitative longitudinal study over 10 years of 24 local government organisations in the UK undergoing radical change.[4] The key concepts included in the model, such as value commitments, interest dissatisfaction, power dependencies, and capacity for action, were assessed at a variety of points in time, enabling the researchers to derive the relationships expressed in them. They found that full organisational "reorientations" occurred when the "value commitments" and "interest dissatisfaction" dimensions resulted in the "precipitation" of a change project, and when certain "enabling dynamics" were also present (expressed in terms of power dependencies and capacity for action). When various conditions in the model diverged from the ideal, the organisational change either never took off (no precipitating dynamics), or was sporadic, oscillating, or temporary (limited enabling dynamics). Clearly, in this case, there are qualitative and linguistic aspects to the modelling process although all of the concepts in the model do have clear definitions and well-defined categorical dimensions enabling a systematic assessment of the validity of the model.

What is the role of language in modelling? Are there qualitative aspects in modelling?

As mentioned above, language is central to modelling in my field. It is through language that concepts and their nuances are defined, and relationships are justified and explained. Since I am a qualitative researcher, many of the models I develop in the course of research are also supported by textual or linguistic data (e.g. quotations from interviews, fieldnotes, documentary evidence) that are brought together to articulate and demonstrate the plausibility of the relations described in models. Models may also of course be supported by grand theory, by logical argument, and in some cases illustrated by examples. All of these involve the use of language.

---

[4]Hinings, C.R. and Greenwood, R. 1988. *The Dynamics of Strategic Change,* Oxford, UK: Blackwell.

How important is notation?

Since models are not always represented mathematically, notation is not as important as in other fields. Note, however, that in Figure 1, the authors label each of the relationships (arrows) in their model with a letter. This letter is then used in the textual explanation to assist the reader in connecting the arguments around specific relationships with the figure.

Models are often said to represent a target system (typically a selected part or aspect of the world). Does this characterisation describe what happens in your field? If so, could you say how a model represents its target? In other words, how do you understand the model–world interface?

Models attempt to represent real-world processes at a level of abstraction high enough that they have some kind of descriptive generality beyond the context of their development.

What is the relation between a model and a theory?

A theory may underpin a model of a more specific phenomenon. Thus, the model presented in Figure 1 is underpinned, among other things, by strands of "institutional theory" which provide a rationale explaining certain relationships in the model. For example, in the abstract of their paper in which the model is presented, Greenwood and Hinings (1996: 1022) note: *This article sets out a framework for understanding organizational changes from the perspective of neo-institutional theory. The principal theoretical issue addressed in the article is the interaction of organizational context and organizational action. The article examines the processes by which individual organizations retain, adopt, and discard templates for organizing given the institutionalised nature of organizational fields.* Institutional theory has been one of the most influential organisational theories in recent years. It focuses on the normative and cultural aspects of organising that are not incorporated into rational economic theories of organisational behaviour, suggesting, for example, that organisations are motivated to take actions that will enhance their legitimacy and that this depends on changing institutionalised norms within their field at a particular time. So, for example, firms in the energy sector in today's world are expected to show concern for environmental impacts and this will affect their propensity to adopt certain internal structures. The argument of the authors that underlies the model in Figure 1 (and specifically relationships a and b) is that the precipitation of organisational change in the direction of adopting new forms depends on both market (or economic) pressures (relationship a) and institutional pressures associated with legitimacy (relationship b), and that these pressures will be manifested respectively through patterns of interest dissatisfaction and value commitments, providing the needed stimulus.

What is the aim/use of the model: e.g. learning/exploration, optimisation/exploitation?

Models in my field are mostly aimed at improving understanding of a phenomenon — i.e. explaining when, how, and why it occurs and/or evolves, rather than in achieving optimisation. Models help us "make sense" of empirical realities (Langley, 1999)[5]. For example, the model in Figure 1 is aimed at explaining when, how, and why radical organisational change does or does not occur. At the same time, some models may help offer guidance for practice. For example, a would-be change agent might deduce from Figure 1 that in order to succeed with change, certain power relations or organisational capabilities might need to be altered.

In case you use computer simulations, what is the relationship between simulations and the model?

I do not use computer simulations. However, other authors in my field have used them. When they do so, the purpose here is to formalise relationships such as those shown in the model above and to explore various contingencies and determine possible outcomes that might be hard to predict in the real world. For example, Sastry[6] developed a simulation model intended to better capture and understand the consequences of a

---

[5]Langley, A. 1999. Strategies for theorizing from process data. *Academy of Management review*, 24(4): 691–710.

[6]Sastry, M. A. 1997. Problems and paradoxes in a model of punctuated organizational change. *Administrative Science Quarterly*, 42(2): 237–275.

so-called "punctuated equilibrium" model of organisational change based on earlier work by Tushman and Romanelli[7] and others. The simulation suggested that organisations are very likely to fail after attempting radical transformations. Sastry introduced various contingencies into the model to examine ways in which this disastrous outcome might be avoided. In general, however, simulation models are rather rare in my field, and somewhat contested because of the inevitable simplifications that they involve, notably in terms of basic assumptions. They often include a number of parameters whose values must be estimated based on relatively little evidence or connection to empirical data. This is not an approach that I favour.

What is a good model?

A good model is a schematic representation and accompanying text that helps explain, understand, and/or "make sense of" a common empirical phenomenon. In my own work, I have referred to three criteria for a good model: accuracy (close fit with empirical data), simplicity (reduction to the most parsimonious expression in terms of concepts and relationships), and generality (with wide applicability). These criteria were originally suggested by Thorngate[8] and later referred to by Karl Weick.[9] Weick noted that these criteria are actually not entirely compatible. As I indicate in my own paper on "Strategies for theorizing from process data,"[10] this means that multiple conceptual models, each positioned differently according to these three criteria, might be equally valuable, but for somewhat different purposes. In some cases, accuracy may be preferable, but this can come at the expense of simplicity and generality, and vice versa.

---

[7]Tushman, M. L. and Romanelli, E. 1985. Organisational evolution: A metamorphosis model of convergence and reorientation. *Research in Organisational Behavior*, 7: 171–222.
[8]Thorngate, W. 1976. Possible limits on a science of social behavior. In J. H. Strickland, F. E. Aboud, and K. J. Gergen (Eds.), *Social psychology in transition*, 121–139. New York: Plenum.
[9]Weick, K. 1979. *The Social Psychology of Organizing*. Reading, MA: Addison-Wesley.
[10]Langley, A. 1999. Strategies for theorizing from process data. *Academy of Management Review*, 24(4): 691–710.

What has been the impact of the development of new technologies or tools (e.g. in cosmology, telescopes, computer simulations for DNA, etc.)?

I am a qualitative researcher. We are seeing developments in qualitative data analysis software that can assist researchers in coding their data and generating conceptual models. However, these technologies are helpful mainly for organising and managing the process, not for generating conceptual models on their own.

How would you define uncertainty? And how does the model help us understand uncertainty?

The model presented in Figure 1 does not really address uncertainty. However, I could see several ways to define such a construct. One way would be to consider "objective" uncertainty, when information about a particular phenomenon is missing. This could be related to the current state of the world, or about the future. A second way of thinking would be to consider "subjective" uncertainty, where a person is not sure about something, or does not know what to do in a particular situation. Such conceptions of uncertainty can appear in conceptual models in my field. For example, a classic contingency model of decision-making by James D. Thompson[11] (latest edition, 2011) considers two dimensions of uncertainty (uncertainty about goals and uncertainty about means), and suggests decision-making strategies that would be appropriate for each of these situations.

How would you define risk? And how does the model help us understand risk?

For me, risk would be defined as the possibility or probability of an adverse event. Risk might be seen as "real" or as socially constructed. Thus, for example, certain scholars in my field, such as Maguire and Hardy (2013),[12] have studied the social, political, and organisational

---

[11]Thompson, J. D. 2011. *Organisations in Action: Social Science Bases of Administrative Theory*. Transaction publishers.
[12]Maguire, S. and Hardy, C. 2013. Organizing processes and the construction of risk: A discursive approach. *Academy of Management Journal*, 56(1): 231–255.

processes and practices by which chemical products are socially constructed as risky.

For you, what is the experience/result in your field that has had the most significant impact? And why?

This is very difficult to say. It depends whether one is interested in impact on practice or impact on other scholars in terms of citations. The most heavily cited papers in my field tend to be theoretical rather than empirical. Those with the most important impact on practice are not always based on empirical research. The most influential model of organisational change in terms of impact on practice is no doubt Kurt Lewin's (1951)[13] three-stage model of "unfreezing," "moving," and "refreezing." This is, however, only indirectly built on empirical research. Its usefulness is based on its metaphorical value in helping managers understand the dynamics of change. In terms of empirical work, Kathleen Eisenhardt is one researcher in the field who has no doubt had a strong impact, because her work on high technology firms is published both in academic journals and in management-oriented journals such as *Harvard Business Review*. Her work has garnered over 100,000 citations in Google Scholar with an h-index of 62. She has published on issues such as strategic decision-making, organisational change, dynamic capabilities, product development, innovation, and boundary spanning. Many of her contributions involve qualitative data and multiple case studies in high-technology organisations. She and her students have been able to achieve substantial breakthroughs in my field of research, as well as acquiring a strong following among practitioners.

---

[13]Lewin, K. 1951. *Field Theory in Social Science: Selected Theoretical Papers.*

Chapter 4

A Conversation with Anthony C. Atkinson

"If your experiment needs statistics, you ought to have done a better experiment."

Ernest Rutherford

Can you describe your field briefly?

I am a statistician with a particular interest in the design of experiments, that is, in the choice of informative conditions at which to make observations.

Present a paradigmatic example of a model in your field, describing it in terms that are accessible to non-experts.

A Clinical Trial on the Treatment of Obesity. Clinical trials are used to assess the efficacy of medical treatments. Often, a new treatment is compared with a standard.

Such websites as those of the National Health Service (NHS) in the United Kingdom and the National Institutes of Health in the United States contain dire warnings about the economic costs of obesity. The costs come from an increased health care burden, poorer work performance, and the need to build physical systems, such as stronger lifts and larger seats in public transport, that can accommodate the needs of the obese. Yet, simple treatments or interventions are often highly cost-effective; examples are programmes for local government employees and in schools to encourage walking and cycling as against driving.

Tsai et al.[1] describe a clinical trial to asses one such simple intervention. The purpose was to compare the effect of standard advice on weight loss with the effect of enhanced counselling in which brief quarterly meetings were enhanced by eight 15–20 minute home visits. The main outcome measured was individual weight loss over the year of the trial. In addition to weight, 14 potential prognostic factors were measured on each participant at the beginning of the trial. These variables included body mass index, diastolic blood pressure, and concentration of low density lipoprotein.

In this trial, there are two treatments, one of which is a standard treatment or placebo. The response of each patient is weight loss over the year of the trial. The simplest model is that the weight loss depends on the treatment, the effect of which is assumed the same for all patients. It also depends linearly on the values of the prognostic factors, the linear relationship being assumed to be the same for all patients. There is also random variability in the measurements of weight loss; the simplest assumption here is that the variance of this variability is the same for all patients.

In symbols, rather than words, this model for the response of patient i, when there are just two prognostic factors, is written as

$$y_i = \delta_i \alpha_1 + (1 - \delta_i)\alpha_2 + \beta_1 x_{1i} + \beta_2 x_{2i} + \varepsilon_i. \tag{1}$$

The treatment effects are α_1 and α_2. The treatment allocated to patient i is indicated by the value of δ_i, which is one for treatment 1 and 0 otherwise. The values of the two prognostic factors for the ith patient are x_{1i} and x_{2i}. The additive independent errors ε_i have constant variance. The parameters α and β have to be estimated from the data. The model is a statistical model because of the inclusion of the error term ε_i, which here is on the

---

[1] Tsai, A. G., Wadden, T. A., Rogers, M. A., Day, S. C., Moore, R. H., and Islam, B. J. 2010. A primary care intervention for weight loss; results of a randomized controlled pilot study. *Obesity*, 18, 1614–1618.

right-hand side of the model, with the observed response y_i, weight loss, on the left-hand side. Because of the assumptions about the errors, the most efficient method of parameter estimation is least squares.

This model may be inadequate. It is possible that the effects of the two prognostic factors are not additive, but that they interact with each other. Then an extra term has to be added to (1) which becomes

$$y_i = \delta_i \alpha_1 + (1 - \delta_i)\alpha_2 + \beta_1 x_{1i} + \beta_2 x_{2i} + \beta_{12} x_{1i} x_{2i} + \varepsilon_i. \quad (2)$$

In some cases, known transformations of the prognostic factors may be needed, such as the logarithm of a concentration. In other cases, the assumptions of the model on the right-hand side, in particular linearity of the effects and errors of constant variance, may be satisfied by some transformation of weight loss, for example, its square root. Determining this transformation may be a matter of data analysis.

The model (1) for the clinical trial is typical of the empirical linear models widely used in statistics. With one prognostic factor, the model would be such that there is a linear relationship between response and the prognostic factor, the fitted relationship having the same slope for the two treatment groups. The treatment effect, if any, will show as a constant difference in response between the two lines. With two prognostic factors, as in (1), the model consists of two planes, rather than lines, but again with a constant difference in response due to the treatment effect.

With the help of this example, could you explain why a model is needed? Could you describe what a model is? And what is the role of mathematics in modelling?

Regression. There is little explicit role of mathematics in the model (1) for the response of the clinical trial. It is just a symbolic representation of the verbal description of the model that precedes it. The hidden role of mathematics in this model is important as a consequence of the additive error term ε_i. The assumptions about the errors determine the way the parameters should be estimated and the estimates interpreted. The usual assumption is that the errors are independent and have constant variance; the responses to the two treatments are equally variable with the variability independent of the values of the prognostic factors. Further, the response of each patient is independent from that of all others. In this familiar linear model (familiar at least to statisticians), least squares provides the most efficient estimates of α and β. Importantly, the assumptions about the errors also lead to inference about an observed value of the treatment effect, that is, the difference in the estimates of α_1 and α_2. The importance of any observed difference depends on the variability in the observations. Here, mathematics was necessary to derive the reference distribution for the observed difference divided by the estimated standard deviation, in this case, Student's t distribution.

Although it is straightforward to write down expressions like (1), the verbal description of more complicated models can become complicated and inefficient. With models that are, for example, the solution of sets of differential equations, mathematical methods are necessary to get from the description of the model (the set of differential equations) to a set of expressions for the observed responses. Whether these are found analytically or numerically, mathematics may be heavily involved.

You have described what you call a "Statistical Model." But not all models are statistical, are they?

Indeed not. Although many models are originally suggested by observation, the focus in developing and exploring non-statistical models is on understanding the underlying relationship between variables observed without error. Concerns about inference and the prediction of a distribution of future results do not usually arise, whereas they do in the analysis of data from the clinical trial. Two examples should make this clear and illuminate some aspects of the process of modelling.

The Ideal Gas Law. The ideal gas law provides an excellent description of the behaviour of many gases under a wide range of conditions. With P denoting pressure, V volume, and n the amount of the gas,

$$PV = nRT, \qquad (3)$$

where T is absolute temperature and the value of the ideal gas constant R depends on the units used. The law was proposed by Clapeyron in 1834 as a combination of Boyle's Law (in which the temperature is constant) and Charles's Law, in which the pressure is constant. As opposed to (1), there is no error term in this model and, indeed, the Wikipedia plot of Boyle's data shows no evidence of departure from a smooth curve, a tribute to the laboratory equipment built by Robert Hooke.

The ideal gas law can also be derived from first principles using the kinetic theory of gases, in which several simplifying assumptions are made, chief among which are that the molecules, or atoms, of the gas possess mass but no significant volume, and undergo only elastic collisions with each other and the sides of the container in which both linear momentum and kinetic energy are conserved.

The ideal gas law neglects both the finite volume of the molecules of gas and the attractive forces between molecules, so departures are to be expected unless the temperature is high and the pressure low. A more general law, due to van der Waals, is

$$\left(P + \frac{\theta_1}{V^2}\right)(V - \theta_2) = nRT. \qquad (4)$$

The values of the quantities θ_1 and θ_2 can be found by theoretical calculation, including the geometry of spherical gas molecules; they are not statistical parameters to be estimated from the data.

An interesting feature of (4) is that it is a cubic equation in V. In statistical work, cubic equations are regarded with suspicion; if a cubic equation is needed, it is taken as evidence that the model is somehow misspecified. This feeling derives from the appreciation of second-order models as being Taylor series approximations to unknown physical laws. If, for example, the maximum of a response surface is asymmetric, requiring cubic terms in the model, a search should be made for a parametric model, perhaps a transformation of the variables, that leads to symmetry and a quadratic model. In the case of (4), Maxwell[2] was motivated by physical evidence of the nature of departures from the cubic isothermal relationship to suggest a linear interpolation of the model for some values of pressure and volume.

---

[2] Maxwell, J. C. 1875. On the dynamical evidence of the molecular constitution of bodies. *Nature*, 11, 357–359.

Chemical Kinetics. The third example combines scientific modelling of the kind associated with the gas law with the statistical modelling of observational error that was the main feature of the model for the clinical trial.

In part of a typical industrial process, chemical A in the presence of a catalyst is converted to a product B. The concentration of A over time is consequently defined by a differential equation. As the reaction continues, the concentration of A approaches an asymptotic value K. This mechanistic model is similar to the ideal gas law in the specification of an understood nonlinear physical law and in the absence of a model for observational error. The purpose of the experiments is to estimate how long the reaction should be run for the concentration of A to be within a specified limit around the asymptotic value. This depends on the values of kinetic parameters in the model which, unlike the universal gas constant R, have to be estimated from the data.

In this simple case, the experimental variables are the initial concentration of A and the times at which observations are taken. If the observations have constant variances, least squares is again the appropriate method of estimation. However, the analysis of the results will depend on which measurements are taken and how they were taken. In the original work of Box and Lucas,[3] it was assumed that independent observations are taken on just one concentration; this corresponds to running the experiment from a starting point to a single specified time and taking one measurement. However, in the study, of which this is a simplified version, each run from a specified starting concentration gives a series of values of the concentration of A and B. The results from different experiments, even under seemingly identical conditions, follow slightly different curves and the observations are correlated. Assuming that the observations are generated by a series of independent errors will give too small an estimate of the variability in the estimated parameters. Consequently, the interval for the estimated time to the target conditions will be too narrow.

---

[3] Box, G. E. P. and Lucas, H. L. 1959. Design of experiments in nonlinear situations. *Biometrika*, 46, 77–90.

In addition to these important departures from the statistical model, there may also be departures from the mechanistic model. Most important of these, in the process under study, was the slow decomposition of B, in an irreversible reaction, to a useless side product. This gradual removal of B from the system slowly reduced the concentration of A, so that the target reduction was achieved slightly earlier than expected. However, the final concentration of product B was lower than would be predicted by the kinetic model.

What constituents besides a mathematical formalism are part of a model?

The specific interpretation of the linear model (1) for clinical trials is completely missing. There are two classes of questions: one derives from the purpose of collecting and analysing the data and the other considers how the data should be collected.

In the majority of clinical trials, inferential interest is focused on the vector parameter α measuring the effect of the treatment. Let α_1 be the value for treated patients and α_2, for the control group. When comparing a new treatment with a standard, interest may be in whether α_1 is greater than α_2. If the new treatment is cheaper or causes fewer side effects than the current standard, then this question may be slightly modified to become whether the new treatment is at least as good as the current one, that is, whether $\alpha_1 - \alpha_2 \geq 0$. Interest in the treatment effect α implies that this effect is the same for all patients; if the treatment is accepted, all patients will receive the same dose.

The parameters β for the prognostic factors describe how the mean response changes with the level of the factor; patients with a higher initial weight can be expected to also have a higher final weight. However, in the model, the difference between the effect of the two treatments for these patients is the same as it is for any other selected set of prognostic factors. The parameters β are not important in themselves and are called "nuisance" parameters. However, they are estimated in the linear model along with α. Many of the measured prognostic factors may, however, have little or no explanatory power. Of the 14 potential variables for inclusion in the model for the obesity trial, only two had any explanatory power. Inclusion of these two variables (one of which was indeed initial weight) in the fitted model produces a more precise estimate of the treatment effect.

The experimental design question is which patients should be allocated to which of the two treatments? With equal variability in the responses to the two treatments, roughly half of the patients should receive one treatment and roughly half the other. This target of equal allocation is often achieved through tossing a fair coin, or its digital equivalent; patients whose toss is a "head" receive one treatment, and patients whose toss is a "tails," the other. An exact division is not crucial. This randomised allocation has the advantage that, on average, it avoids two kinds of biases. One comes from an unequal allocation of treatments over prognostic factors; for example, if age is an important prognostic factor, a similar distribution of age for the two treatments is desirable. A second kind of bias is due to factors that are not included in the model, because they cannot be measured before the treatment allocation is made. In the trial on obesity, the home visits were made by members of a team of medical assistants. It can be expected that there will be specific effects associated with individual team members. The randomised allocation, on average, also provides security against biases from such sources.

In the trial on obesity, the patients clearly knew which of the treatments they were receiving. However, where possible, further biases can be avoided by "blinding" the trial, that is ensuring that the patients do not know whether they are receiving a standard or an experimental treatment. As an example, Tamura et al.[4] assessed the use of a drug, fluoxetine, in the treatment of 88 patients with depression. It is to be expected that the attention and effort of being involved in a clinical trial will help lift depression. To allow for this potential effect, fluoxetine was compared with a placebo — an apparently similar treatment with no physiological effect. In this trial, the allocation of patients to treatments was again randomised.

Provided the allocation from the randomisation is strictly adhered to, it can avoid "selection bias," that is the well-documented tendency of

---

[4] Tamura, R. N., Faries, D. E., Andersen, J. S., and Heiligenstein, J. H. 1994. A case study of an adaptive clinical trial in the treatment of out-patients with depressive disorder. *Journal of the American Statistical Association*, 89, 768–776.

doctors to allocate the treatment they think most suitable for specific patients. There are, however, further, more subtle, forms of biases. Patients are more likely to respond better to a new treatment than a standard one or placebo, provided they know which they are receiving. Similarly, doctors and those assessing the patients' performance may be biased in favour of a particular treatment and convey this to the patient, even by the tone of voice in which questions about the response to treatment are asked. For this reason, "double blind" trials in which neither patients nor the medical staff know which treatment has been allocated are the standard.

It is clear that the description and understanding of the clinical trial represented by (1) requires more than just the bare postulation of the algebraic form. However, this form can be used to examine important aspects of trial design and consequential inference, such as various forms of randomisation and their effects on the precision of estimation and on the extent of selection bias.

How important is notation in modelling?

Clear notation is a great benefit in any mathematical work, helping more rapid development of the argument and, with luck and good judgment, showing the relationship with other mathematical structures.

It also seems to be extremely important in communication. Equation (1) is written in a notation that is mostly standard to statisticians. Responses are customarily denoted y, explanatory variables and factors, x, and the regression coefficients, β. The labelling of the treatment effects by α is also standard. In theory, any sets of letters could be used, but particular meaning is applied to these symbols; Greek letters are used for parameters and explanatory variables are customarily labelled with letters at the end of the Latin alphabet. Trying to work with a different set of symbols is possible for individuals. But, for ease of communication between people, standard notation is a great blessing. An example, not from modelling but computing, is the widespread convention in the Fortran language that integers, unless otherwise defined, begin with the letters I–N, although more recent languages have disregarded this convention.

What is the role of language in modelling? Are there qualitative aspects in modelling?

There is little role for language in the models considered here. In the clinical trial on depression, the responses may well be answers to a battery of questions. Otherwise, the obesity trial and the chemical kinetic model describe systems in which language has no role. Although, as I say, the symbols attached to quantities are important for communication.

Language is, however, important in communicating the output of a model. This is particularly important in understanding some basic, but important, statistical ideas, especially the intervals that are often quoted around estimates of parameters like α_1 and α_2 in (1). Are these non-varying physical constants like the R in the ideal gas law? In that case, a confidence interval around an estimated value is a statement of uncertainty due to measurements. But the parameters in the model for the clinical trial might be expected to vary from patient to patient. The parameter for each patient can then be modelled as realisations of random variables, the prior distribution of which can be modelled, perhaps on the basis of subjective prior information. In that case, a Bayesian interval would be a reflection of a prior estimate of the population variability of the parameters modified by the data.

Models are often said to represent a target system (typically a selected part or aspect of the world). Does this characterisation describe what happens in your field? If so, could you say how a model represents its target? In other words, how do you understand the model–world interface?

The purpose of a clinical trial is to decide the "best" treatment in the cheapest way, where best may include costs, absence of side effects, and ease of deliverability — a course of pills is more likely to be followed correctly than a course of injections requiring attendance at a clinic. However, the statistical modelling and analysis are only a small part of the logistics and planning of a clinical trial and the statistical model for treatment allocation could well be incorporated in a larger model for patient recruitment and resource allocation.[5]

---

[5] Anisimov, V. V. and Fedorov, V. V. 2007. Modelling, prediction and adaptive adjustment of recruitment in multicentre trials. *Statistics in Medicine*, 26, 4958–4975.

What is the relation between a model and a theory?

There is little relationship between the empirical model for a clinical trial and theory, at least in any way that is capable of precise mathematical formulation. It is expected that, on average, more attention will improve self-discipline and the ability to eat less and exercise more. But the question of interest was, by how much? There is more theory in the mechanistic model for chemical kinetics. Although the simple model is known, departures such as the production of by-products, if any, can be included in a form using kinetic theory. On the other hand, the ideal gas law can be completely derived from theory. The aim of early experiments was to verify the relationship, which led to theoretically-based extensions of the model to allow for extreme conditions. Here, interest is overwhelmingly in verifying a theory. It is this kind of experiment that was of interest to Rutherford; in his case, to model the structure of atoms.

As I have already mentioned, the theory of statistical inference is crucial to the use that is to be made of statistical models where the purpose is not just to explain what is going on, but to quantify the uncertainty in any statements caused by observational error.

What is the aim/use of a model: e.g. learning/exploration, optimisation/exploitation?

In the clinical trial on obesity, the ultimate aim is to make the decision as to which treatment is more cost-effective. In a more typical clinical trial in which several drugs, or doses of the same drug, are compared, much has to be learnt about the dose to be given and the nature of any side effects. The dose level and frequency need to be optimised and a model is helpful in learning, in the design of efficient experiments, and for optimisation.

The aim of the model in the kinetic example is similar: a series of models may be necessary until one is found that incorporates all important features. Once this learning phase is over, the model can be incorporated into an optimisation of the process conditions and statements made about the distribution of yield and hence of profit.

In the case of such nonlinear models, it is virtually impossible to design even halfway decent experiments, let alone efficient ones, for choosing between models and estimating parameters without making use of tentative models. I talk a bit about model-based optimum experimental design at the end of this conversation.

In case you use computer simulations, what is the relationship between simulations and the model?

One use of computer simulations in the model described here for the clinical trial is to compare designs for treatment allocation when the treatments are allocated sequentially in the presence of prognostic factors. The purposes of the design are to improve efficiency by trying to ensure that the patients allocated to each treatment are representative of the population of potential patients; in many trials, this balance needs to hold whenever the trial is stopped. At the same time, selection bias needs to be kept low. Although there are some theoretical results for large trials, for smaller trials, simulation is necessary to elucidate trial properties.

Simulations may well be performed using a simplified model. In a complicated nonlinear model, sensitivity analysis[6] can be used to eliminate those parts of the model which have a negligible effect on responses of interest over the range of conditions to be studied.

This simplification is distinct from the treatment of complex models in, for example, engineering, where the systems may be described by large linked sets of nonlinear differential equations. A single simulation under fixed conditions from such models may take one day or more. To explore the effect of changes of factor levels on the outputs, simulations are run at a few carefully chosen conditions and a smooth interpolation model is used to provide estimates of the response at arbitrary factor settings. Here, there are two levels of the model: the differential equations and the interpolation approximation. The design of such computer experiments is described by Santner *et al.*[7]

---

[6] Saltelli, A., Ratto, M., Andres, T. Campolongo, F., Cariboni, J. Gatelli, D., Saisana, M., and Tarantola, S. 2008. Global Sensitivity Analysis: The Primer. Wiley, New York.
[7] Santner, T. J., Williams, B., and Notz, W. 2003. *The Design and Analysis of Computer Experiments*. Springer–Verlag, New York.

What has been the impact of the development of new tools in modelling?

I don't think there has been any dramatic step change in modelling. It becomes increasingly possible to make better models of systems of increasing complexity. These models can routinely include objective or subjective prior information and can then be tested against data in a variety of ways using robust and diagnostic techniques. Linked graphical displays provide keen insights into the dependence of models on aspects of the data to which they are fitted. And, of course, datasets become increasingly large. But probably the most dramatic development in modelling in the next few years will be virtual reality displays of the output of models, rather than aspects of the modelling process itself.

What is a good model?

A good model includes all important features without being over-elaborate. In the case of the clinical trial model, all prognostic factors that affect the response need to be included. Inclusion of other factors that do not have an effect leads to less precise modelling, as the parameters associated with the effects still have to be estimated.

Of course, once the data have been collected, tests can be used to remove unnecessary factors from the model and so improve precision of predictions and inference. In the trial on obesity, the number of variables was reduced from 14 to two in the analysis stage. However, at the design stage, all variables were included. In general, seeking balance over too many factors may lead to unnecessary imbalance once the superfluous prognostic factors are removed from the model.

How would you define uncertainty? And how does the model help us understand uncertainty?

There are probably three main sources of uncertainty. Given that the statistical model is correct and the parameters are known without error, there is variability in the observations; it is not possible to predict with certainty what response a patient will have to the treatment. Even when all relevant factors have been measured, there remains a distribution of the response, something that is at odds with Rutherford's statement quoted at the beginning.

The second source of uncertainty comes from estimation of the parameters. This introduces uncertainty into the statement that one treatment is better than the other. The average difference can be estimated from the data, but this estimate will itself have a distribution; the uncertainty resulting from this distribution needs to be stated, perhaps as a confidence interval, in any reporting of the results.

The third source of uncertainty is that coming from the correctness, or otherwise, of the model. It may well be that, for example, certain identifiable subsets of patients respond differently from the others. In that case, the model needs to be extended to cover these groups. Not identifying them leads to models that are unnecessarily imprecise. Of course, much more important is the aspect of personalised medicine, where the attempt is made for each patient to receive the treatment that is best for that individual.

How would you define risk? And how does the model help us understand risk?

In the clinical trial example, there are two risks associated with the statistical analysis. One is that the trial will fail to identify the superior properties of a new treatment, often because the trial is not large enough. Such trials are referred to as "underpowered." The reverse risk is that the trial may be excessively large because the difference between the treatments cannot be established. These are both somewhat technical forms of risk, but the failure to identify a useful molecule is likely to mean that the line of research may be dropped and that related potentially useful products will not be explored.

Of course, there are more obvious forms of risk involved in clinical trials. Senn *et al.*[8] describe a first-in-man trial in which six participants nearly lost their lives. This is a very real, if non-technical, form of risk which could have been avoided if the current best procedure had been followed. In particular, the treatments were allocated to all patients at the same time. Instead, they should have been allocated sequentially to indi-

---

[8] Senn, S., Amin, D., Bailey, R. A., Bird, S. M., Bogacka, B., Colman, P., Garrett, A., Grieve, A., and Lachmann, P. 2007. Statistical issues in first-in-man studies. *Journal of the Royal Statistical Society, Series A*, 170, 517–579.

vidual patients, allowing sufficient time to observe the unexpected and severe adverse reactions that occurred.

What do you consider to be the work/result that has had the most significant impact on your field? And why?

Optimum Experimental Design. The methods of optimum experimental design, first formalised by Kiefer,[9] are essential for the efficient design of experiments in anything but the simplest cases. For the clinical trial, they provide methods of treatment allocation in the presence of prognostic factors and provide a standard for comparison of the efficiencies of other procedures. Likewise, they provide methods for experimental design for the kinetic model in which the parameters are entered nonlinearly.

Application of these methods requires

- a design region;
- a model;
- a design criterion.

The design region describes the set of conditions that can be chosen for each observation. In the clinical trials, each patient can be given either placebo or treatment, so there are just two points in the design region. The model is the linear model (1), with additive errors of constant variance. Often, the interest is in comparing the two treatments when a natural design criterion is to minimise the variance of the estimated difference $\hat{\alpha}_1 - \hat{\alpha}_2$.

The design for the clinical trial is robust to some departures from the model (1). One example would be an interaction between the two prognostic factors, adding the extra term included in the model in (2). The design is also robust to some departures from the assumption that the errors have constant variance. For example, if the response were a survival

---

[9] Kiefer, J. 1959. Optimum experimental designs (with discussion). *Journal of the Royal Statistical Society B*, 21, 272–319.

time, taking the logged time often gives a response with errors of nearly constant variance.

A mathematically formulated model is hardly necessary for the design of the simplest clinical trial in which there are two treatments to which patients are randomised in the absence of prognostic factors. Simple procedures are also possible if there are only a few prognostic factors which are categorical. In the fluoxetine trial,[10] two binary prognostic factors were included in the design leading to a 2^2 factorial arrangement of these factors. Treatments are then allocated at random within each of the four strata defined by the values of the prognostic factors for each individual. However, in the clinical trial on obesity, there were 14 potential prognostic factors and only 50 patients. On average in this clinical trial, simple randomisation to treatment can produce allocations so unbalanced across prognostic factors that the information lost, compared to an optimum design, is roughly equivalent to the measurements from 10 patients.

In fact, both trials were sequential so that more sophisticated methods of treatment allocation were necessary. In both cases, it was not known how many patients would be recruited before the trial terminated. Accordingly, the treatment allocation had to provide some randomisation against bias combined with an extent of balance over the prognostic factors throughout the trial. A further enrichment of the design problem in the fluoxetine trial was that a response from each patient was available shortly after treatment. As evidence started to become available as to which treatment was better, the allocation probability was changed adaptively to provide patients with a higher probability of being allocated the better treatment. However, some randomisation was still required. The methods of optimum design suggest efficient rules for such sequential adaptive designs. Simulation based on the models allows assessment of the properties of a variety of rules, leading to an appropriate trade-off between loss of information due to imbalances in allocation and biases from being able to guess which treatment will be allocated next.

These considerations are rather different from those in the design of the chemical experiment where choice of initial concentration and observation times depend strongly on the model and the values of the

---

[10] Tamura, R. N. *et al.* 1994, *op cit.*

parameters; it is almost impossible to guess an efficient experiment and increasingly so when the temperature of the reaction is allowed to change over time. There is no sensible alternative to the use of a model-based optimum design.

Literature

Model-based optimum experimental design of the kind sketched here was developed by Kiefer.[11] Kiefer's papers on experimental design are collected in Brown et al.,[12] which also includes a survey of Kiefer's work by Wynn.[13] An introduction to optimum design, with some emphasis on the process industries, is in Atkinson et al.,[14] who include material on designs for the detection of model inadequacies. Atkinson and Biswas[15] describe the use of optimum design methods in sequential and adaptive clinical trials. Box and Lucas[16] introduced the methods into design for nonlinear models arising in chemical kinetics. Two more recent book-length treatments of optimum design with an emphasis on nonlinear models are Pronzato and Pázman[17] and Fedorov and Leonov.[18]

---

[11] Kiefer, J. 1959, *op cit.*

[12] Brown, L. D., Olkin, I., Sacks, J., and Wynn, H. P., editors, 1985. *Jack Carl Kiefer Collected Papers III*, New York. Wiley.

[13] Wynn, H. P. 1985. Jack Kiefer's contributions to experimental design. In L. D. Brown, I. Olkin, J. Sacks, and H. P. Wynn, editors, *Jack Carl Kiefer Collected Papers III*, pages xvii–xxiv. Wiley, New York.

[14] Atkinson, A. C., Donev, A. N., and Tobias, R. D. 2007. *Optimum Experimental Designs, with SAS*. Oxford University Press, Oxford.

[15] Atkinson, A. C. and Biswas, A. 2014. *Randomised Response-Adaptive Designs in Clinical Trials*. Chapman and Hall/ CRC Press, Boca Raton.

[16] Box, G. E. P. and Lucas, H. L. 1959, *op cit.*

[17] Pronzato, L. and Pázman, A. 2013. *Design of Experiments in Nonlinear Models*. Springer, New York.

[18] Fedorov, V. V. and Leonov, S. L. 2014. *Optimal Design for Nonlinear Response Models*. Chapman and Hall/ CRC Press, Boca Raton.

Chapter 5
A Conversation with Arthur Jaffe

Can you describe your field briefly?

My field is to understand Nature as mathematics. What that means is, we try to understand the world through a logical (mathematical) description of physical phenomena, one that also has the power of prediction.

My own special interest has for many years focused on the two biggest achievements of 20th century physics — relativity and quantum theory. One could regard these subjects as the elephants in the room. Finding each of them revolutionised our understanding of Nature, and physics has focused on them ever since their discovery. Relativity deals mainly with the macro-world of classical physics and the cosmos. Quantum theory deals mainly with the micro-world. It not only frames our understanding of elementary particles, atoms, and molecules, but it also provides the framework for these objects to be the building blocks of gases, liquids, and solids — namely, all of Nature. Both relativity and quantum theory have had enormous practical importance in our lives.

Yet, after almost one hundred years of studying relativity and quantum theory separately, we still do not understand whether these two concepts are *mathematically* compatible with one another! This leads us to ask an even more fundamental question: is it possible to describe Nature through mathematics and to understand both of these phenomena? The alternative is we can access Nature only through understanding different and imperfect models whose domain of relevance may not overlap.

Among the general public, most persons take it for granted that physics is an all-encompassing and fundamental science, actually part of mathematics. However, in the field, experts harbour a lurking doubt whether we can find a model that both describes physical phenomena in detail, and also meets the standards of mathematics.

Present a paradigmatic example of a model in your field, describing it in terms that are accessible to non-experts.

One of the simplest and oldest models forms the starting point of most investigations in mathematical physics. It is the mathematical representation of space and time in the universe by a continuum of points. It provides the fundamental arena in which we describe our world. This picture permeates Euclidean geometry that we learn in school; it is the foundation of Newton's calculus; it forms the basis for Maxwell's theory of electromagnetism; and it remains the basis for modern developments — including relativity and quantum theory. We really do not know whether this space–time continuum is a correct model, yet it forms the basis for most of theoretical physics.

A more complicated model, based on this continuum, is to assume that space–time is compatible with relativity. For special relativity, this means that space–time is the space described by Minkowski, which provides a model space–time that appears to allow for a description of elementary particles.

We'll come back to this a bit later, but would you say you use models mainly to describe Nature?

Absolutely. Models that describe Nature are the basis for all my work. The idea of such a model is to attempt to isolate a part of physics in a self-contained way, so one can describe completely an idealisation of Nature which is a subset of the world. Such models underlie all of physics; by extension, they also include models for chemistry, for engineering, or even for finance. And when one considers such models to describe Nature, they should be logically sound — even though they may be applicable only in special realms.

So, models describe the state of Nature and also movements or evolution?

Evolution is a central feature of Nature, so evolution in time needs to play a central role in a model of Nature. Often, the model arises as a mathematical equation that describes the time evolution of Nature. This is the case in quantum theory, where the equations of Schrödinger, Heisenberg, and Dirac play that role. These equations can come in many different forms, and finding the appropriate form of the equation comes down to discovering a "law of physics." This provides the model.

However, such a dream is grandiose, for one must also show that the equation one conjectures makes sense as mathematics. In the case of planetary motion, Newton had to invent calculus to do that. For the problems with quantum theory and special relativity, one also needs to invent new mathematics in order to show that the equations themselves are mathematically consistent. This is what people in physics try to do: they try to find the equations of evolution, and by doing that, they discover the constituents of Nature. They make predictions based on the model. However, the key conceptual point is that they also need to show that the model fits into mathematics: old or new.

So, with the help of this example, could you explain why a model is needed, and what it is?

A model is needed because we want to translate the world into a set of mathematical statements or equations. In order to describe something by an equation mathematically, you need to have some idea of what the symbols mean. And the model provides the idea of what the symbols mean. Thus, the model may contain the notion of "particles" as something that one can derive from the equations. And forces between particles should also arise as a consequence of finding the right equations (model).

So, my next question, which is: what is the role of mathematics, I think is quite essential from what you are saying?

Everything in theoretical physics revolves around mathematics. This really means that one has a logical framework. Mathematics isn't fixed — mathematics keeps changing as one discovers new ideas. And by inventing a new model, it may be necessary to invent new mathematics in order to understand it. And that's been the history of this subject — that by understanding Nature, we actually discover new mathematics. For example, understanding quantum theory without relativity led to many new insights into practical problems about the real world. But it also led to new insights and theories about differential equations, analysis, probability theory, algebra, representation theory, and geometry. The new model of quantum theory also revolutionised mathematics!

So, is it like inventing new tools to describe something? Or, it's more than tools, isn't it?

Yes. You can think of starting with certain primitive tools, but then refining your tools as you use them. There is a back-and-forth between the discovery of the model, the development of the tools, and the understanding of the model. The tools can be used to understand things, but then have a life of their own — they have their own families. They have their children and they go on merrily; but sometimes, these children turn out to be related to things back in the real world that you never imagined when starting out!

So, a naïve question: would you say that sometimes mathematics, this new mathematics, will describe or represent more than what you originally intended it to describe or to represent, and then it makes you understand new things from Nature?

Absolutely. And in fact historically, when quantum theory was invented, the modern quantum theory started in the 1920s with the works of Jordan and Heisenberg, and also Dirac, but the mathematics associated with that became so complicated that people in the 1950s began to think that maybe it was not even possible to describe Nature by mathematics. Because the theory appeared so complicated, they came to believe that it was impossible to get to the end. And so the mathematics which had been tied to physics historically in the last century seemed to divorce physics. And now we're trying to get them back together again.

I like your way of presenting these things — people might be surprised.

Once new tools were discovered, it turned out that the tools themselves have a much wider validity. The tools can be used in engineering, economics, medicine, and so forth — in ways unimaginable at the time the tools were invented. This is an old story, which I wrote about some time ago in an essay "Ordering the Universe: the Role of Mathematics." Let me quote a bit from the essay about Fourier analysis:

> In the early 1800s, Jean Baptiste Joseph Fourier, newly returned from his post as civil governor of Napoleonic Egypt, set out to understand the problem of heat conduction. Given the initial temperature at all points of a region, he asked, how will heat diffuse over the course of time? It was curiosity about such phenomena as atmospheric temperature and climate that led Fourier to pose the abstract question. In order to solve the heat diffusion equation, Fourier devised a simple — but brilliant — mathematical technique. This equation turned out to be easy to solve if the initial heat distribution was oscillatory — that is, essentially a sine wave. To take advantage of this, Fourier proposed decomposing any initial heat distribution into a (possibly infinite) sum of sine waves and then solving each of these simpler problems. The solution to the general problem could then be obtained by adding up the solutions for each of the oscillatory components, called harmonics.
>
> French mathematicians, such as Lagrange, sharply rejected the idea, doubting that these simple harmonics could adequately express all possible functions, and cast aspersions upon Fourier's rigor. These attacks dogged Fourier for two decades, during which he carried his research forward with remarkable insight. Today we owe an enormous debt to his remarkable tenacity, his stubbornness, and his ability to proceed in spite of formidable doubts in the minds of the leaders of the scientific establishment. Fourier found it difficult to publish his work even after he received the 1811 grand prize in mathematics from the Académie des Sciences for his essay on the problem of heat conduction, because the academy's announcement of the award expressed grave reservations concerning the generality and rigor of Fourier's method. Fourier persevered and finally his work won general acceptance with the publication of his now-classic *The Analytic Theory of Heat*, in 1822.

The method of harmonic analysis, or Fourier analysis, has turned out to be tremendously important in virtually every area of mathematics and physical science, much more important than the solution of the problem of heat diffusion. In mathematics, it has become a subject by itself. But in addition, the theories of differential equations, of group theory, of probability, of statistics, of geometry, of number theory, to mention a few, all use Fourier's technique for decomposing functions into their fundamental frequencies. In physics, engineering, and computer science, the effect has been no less profound. At least as important as the numerous applications to science and engineering has been the application of Fourier analysis to mathematics itself. Like other scientists, mathematicians are constantly searching for new tools to solve their theoretical problems. Frequently, it happens that techniques discovered to solve one abstract problem later apply to a wide variety of others.

If you need to be convinced of this, look under "Fourier" in the catalogue of a university science library. At Harvard's, for example, there are 212 entries, of which the first 10 are Fourier analysis in probability theory, Fourier analysis in several complex variables, Fourier analysis of time series, Fourier analysis of unbounded measures on locally compact abelian groups, Fourier analysis on groups and partial wave analysis, Fourier analysis of local fields, Fourier analysis of matrix spaces, Fourier coefficients of automorphic forms, the Fourier integral and its applications, and Fourier integral operators and partial differential equations.

So, besides this mathematical world, what do you have in a model? Would you say that you have something else besides mathematics?

If you want the model to reflect the world around us, you have to have the concepts that we see in the world around us as part of the model. So, I gave you a very simple example before — points in space–time — but you want to be able to construct objects, and other things coming out of your picture, your model. So in fact, the model that I was talking about, the consistency of relativity and quantum theory, these are based on descriptions where we would like to predict particles that exist and have been seen at

CERN and in other places in laboratories, and the forces between them, based on some simple principles.

Are these principles the results of some other models? Or are they based on some evidence?

The history of models about Nature has to be hierarchical; it needs either to include what came before, or else a new model needs to incorporate the earlier success of models in the same area. One should be able to reproduce what has been understood in the past.

What is the role of language in modelling? Does it have any role at all?

Everything has to be framed in language. And in fact, mathematics is the language of the models I use. It's hard to separate the two. And when you get into these questions very deeply, you're forced immediately into questions of philosophy, and so it becomes quite complicated. It is true that sometimes complicated or catchy names bring attention or notoriety to a particular model. But this "marketing" is something I try to avoid.

So do you have any more qualitative aspects in your modelling? I guess like the description of what you mentioned, these particles — I guess there is a descriptive part, a bit more qualitative?

Well there was a famous lecture given in the 1960s by Mark Kac. It arose from the coincidence of geometry and analysis. The theme was what you could learn from knowing the frequencies you hear from a drum. The question is whether you could actually tell the shape of the instrument by listening to it carefully, so Kac asked, 'Can you hear the shape of a drum?' This became a very famous question, because of course you can find out certain things from hearing certain tones coming out of the drum, but people wondered whether you can describe the shape of the drum completely. It turns out that the answer is no. But it took a long time to find that out. You can find the size, you can find the perimeter, and you can find out many things about the drum; but you can't completely determine its shape. And now we have, coming out of physics, another mathematician, Alain Connes, who invented a new type of "non-commutative" geometry. We don't know if it really applies to Nature, but certainly it would be beautiful to build quantum theory into geometry. In any case, he asked, "Can you hear non-commutative geometry?" While we don't yet have the answer to that, these qualitative questions linger. Such theoretical questions go beyond the more detailed things that a scientist might measure in a laboratory; they ask about what you *are able to* find out, rather than details of what you *actually* observe.

How important is notation in modelling?

Notation can be very important, both in its aid to identifying and relating to the model, and also in emphasising and encapsulating the simplicity of a model.

Coming back to what models are used for: models are often said to represent a target system, some aspect of the world. Does this characterisation describe what happens in your field?

Well yes, and this is actually part of the big question because historically in my field people tried to isolate a small part of Nature and to describe it mathematically. This goes back to Newton's description of planetary motion, or to Maxwell's description of electromagnetism, or to Boltzmann's description of statistical physics, or Gibbs' description of statistical physics, and Einstein's description of relativity, or Schrödinger and Heisenberg's description of quantum theory. They all isolate some small part or phenomenon that can be described very well. But when you try to put these things together, the situation gets much more complicated. And we don't know, for instance, if electromagnetism interacting with matter can be described completely mathematically within quantum theory. This is really a big question in physics. It is the question of how within that field a small part of Nature can be isolated and described. For that is the basis of a model.

In fact, we think that probably the answer to the question about electromagnetism is "no." And that drives people a little bit crazy, because they think they have to include more and more and more. But then the theory gets more and more complicated and the model gets more and more complicated, and the understanding of it mathematically also gets more complicated. So in the end, we don't know if you can make it all work!

Would you say somehow it's easier to have a global model, but if it becomes really complicated, then you have to slice or focus on certain aspects of the world?

No, what I'm trying to say is, if you have a global model, it becomes very difficult to understand how it all fits together, to understand the whole thing, in fact, it challenges the human mind to try and do it. And if you don't have a global model, then even simple things that you want to put together don't seem to fit. So you're in a conundrum.

So it's a bit like climate and weather?

Yes, it's like climate and weather, where you believe you know the equations, but you can't solve them; and you model them on a computer, but you don't really trust the solutions. You don't know how much data to put in to predict the future.

We will come back to this a bit later. How do you understand the model–world interface?

Oh I don't know that I understand that really well; that's the experimental part of science and I'm not an expert on that at all, so I really can't comment very much about it in detail. But I do know that the most accurate measurements that are made in the world have to do with the 13 decimal point accuracy to which one has measured the magnetic moment of the electron — namely, the reaction of the electron when one puts it in a magnetic field. And this experiment agrees down to the last decimal place with rules that physicists devised to calculate the number. In fact, these rules are extremely complicated and took some 60 years to refine and work out. Yet, we do not know that these incredible triumphs of experimental and theoretical science can be backed up by a mathematical theory, like those theories of classical physics that we know from history.

What is the relationship between a model and a theory?

Well, the model is what you have to base your theory on. Models have to be simple, elegant, beautiful, in order for them to be appealing. But once they're appealing, you don't know if they really fit together and work, and so that's where the mathematics comes in, you have to turn the model into mathematics, and then it becomes a theory.

So you would say a model is more a relationship between different constituents, but it can be a sort of logical relationship?

I would say a model is a hypothesis, and the theory is like a mathematical proof.

So the model is like a conjecture, and then...?

It's a conjecture, and a theory is some proof that it fits together and maybe also that it applies to the world.

How do people come up with the conjectures? Are they based on existing theory and somehow you go back up to this upper level of conjecture? How do people come up with the conjecture?

You have to be...it takes a certain genius. You have to have bright eyes, and some nice ideas, and then you have to take into account experience and where you're aiming, that's the way you do it.

And then the theory: the theory proves that the conjecture is right, that it's actually working? It leads to another conjecture I would say?

Of course, yes. But let me tell you a little story: when I was a student, I heard a lecture by Paul Dirac. He was very important in the early development of quantum theory, and he gave a lecture in Princeton. He talked about the history of the relativistic equation for the electron, which most people call the "Dirac equation," although he never used his own name in speaking about things. In any case, Dirac gave this beautiful lecture at the Institute for Advanced Study, in the seminar of Robert Oppenheimer, who said: 'Professor Dirac, at what point did you realise that your equation was correct?' And Dirac replied (I'll say it in the first person), 'I was working at home and had to go to the library in order to look up the spectrum of hydrogen. But, as I was getting my bicycle out of the garage, I realised that this equation was so beautiful it had to be right!'

Ah! I think we will come back to this when we have the question about what is a good model, because this is so beautiful as an answer! It's very interesting, and quite different from what other experts in other fields have said about models.

What is the aim and use of a model? Here I have mentioned a few things, but it might be completely different. So learning, exploration, optimisation, exploitation? It really depends I guess on the field. In your field what would you say?

I think the use of the model is to have something specific to focus on, and then to pose a question that might have a concrete answer. With a model, one does not need to focus on Hilbert's question 'Can one axiomatise physics?' Rather, one can ask whether a particular model of a part of physics is relevant.

So the aim is really to have a sort of representation or a better understanding of how things work?

Yes, to have some conjecture of how things work.

Do you use any computer simulations — or are computer simulations used in your field, not necessarily by you, but by others?

Generally, I don't use computer simulations in my own work. Occasionally, I use them to test some idea, in order to see if it could possibly be right. In this case, the computer simulation gives a first idea, and then you try, based on the outcome of the simulation, to actually prove it mathematically. My own use of simulations is very elementary and limited.

But other people use computer simulations to test ideas, and some people even use computer simulations to try to formulate new ideas, or to actually find approximate solutions to equations. Some people even use computers to test mathematical theories: in geometry or in number theory. In the latter, one has used a computer to locate many millions of zeros of the Riemann zeta function. The answers all fit into the conjecture of Riemann; but, while this leads to some degree of confidence, we do not know that this (possibly the most famous unsolved) mathematical conjecture is true. And only with a true mathematical proof would the many consequences of the conjecture be valid. This is nicely depicted in a book *Prime Obsession* by John Derbyshire.

Coming back to what you said earlier about model and theory, this will be for the theory part, more for like the proof in the understanding?

It's the part that I'm interested in, the proof, yes.

So what is a good model? We mentioned this a little bit earlier.

Well, as we said before, a good model has to be intuitive, it has to be pretty, it should be possible to encapsulate it simply, although understanding it may be extremely complicated. Again, I can quote something from mathematics: Fermat's Theorem is a very simple mathematical theorem that any high school student can understand, and yet it took 350 years for mathematicians to show that it was right.

So this was really the first part, on the modelling side. And now the second part is about risk and uncertainty, which may also be very different in your field and your experience. How would you define uncertainty?

Well, I think there are different types of uncertainty. First, when you're dealing with human beings, you can always make errors, everybody at some point makes mistakes, even computers make mistakes as well. So you have built-in the fact that certainty isn't guaranteed in our life. On the other hand, mathematics probably is more certain than anything else in science. So we believe that the degree of certainty — you have to talk about degree of certainty and degree of risk — and the risk is rather low if you check something very carefully, and multiple people check it, that it's not correct. But there's another type of uncertainty, and that's the uncertainty of what the human mind can do, or what the mathematical theory can do, and we never know, for instance, if you have a model or a hypothesis, whether you can really show if it's either right or wrong. And that's another type of uncertainty: you might spend your whole lifetime trying to show that a certain model is right, and never achieve it; and you don't even know if you can achieve it. So that's another type of risk, which you have to manage, mainly by having good intuition about choosing good things to work on.

Do you make a distinction between uncertainty and risk? Or not really?

No, not really. While they're certainly different, they're so tied up together that their practical consequences can be the same. Offhand, I haven't thought that one through.

That's fine. For economists, for instance, there is a huge difference, but it really depends on the field; and I guess in your field, because as you said, mathematics is somehow really clean in terms of risk associated with it, so…

Risk, I think, is involved in going off in the wrong direction, in a tangent. And there it's not a question of uncertainty, but the risk that you're not going in a fruitful direction.

And so it seems, it's interesting — the only expert I can relate with at this stage is the cosmologist I interviewed, because for him the huge progress in his field was related to the development of tools, observation tools and things like this. So it seems that in your field, the development of mathematics and mathematical areas have a huge impact on what can be done. If it's not mature enough, then people will not be able to make progress during their lifetime, so we need to wait several generations.

Well, in my generation, something which happened which is very important and changed the way that everybody thought about these subjects is that it used to be, maybe 60 years ago, that people were extremely specialised, and you developed A, B, C, D, or E. And then people started to discover relationships between A and B, B and C, D and E, and now more or less everything seems to be related. And that makes it very much more interesting, and also very much more difficult, because instead of specialising and becoming an expert, you have to have some knowledge across many different fields in order to be able to do that. That makes the work so much more difficult. Not only in mathematics has this connection come about, but between mathematics and Nature, between mathematics and physics. So within physics, in fields of physics that people thought were separate, as I mentioned before — like statistical physics, quantum mechanics, classical mechanics — they all now can be related by certain ideas, and in order to see these relations, you also need different aspects of mathematics that seemed before to apply one to one, another to another, but now they are all connected. So this makes everything exciting. That's a big change that's happened during my lifetime.

Did you find it difficult to interact with people at the beginning who had a different expertise, still in physics, but different fields of physics? Since now things are more interdisciplinary within physics — do you have any communication problems, or any difficulties at this stage?

Yes, there are major difficulties, and the difficulty comes because now there are so many more scientists and so many more mathematicians, so each individual scientist or mathematician is surrounded by a cloud of colleagues and students, and that makes it very much more difficult to interact with people in different fields.

I was thinking more in terms of language? Maybe what people mean by certain words, or by certain concepts?

Well, sociologically people become isolated in certain groups, and so even though interaction should take place more and more, however, from a practical side it becomes more difficult.

The last question, which is quite a personal question: what is the experience or result in your field (it can be one of your own, or it can be something also more general) that has had the most significant impact on you, and why?

Well, I think it is what I just explained — that everything has become interconnected, and this has become very important for me, because I was involved in making some of the connections, but I could see everywhere in the field this was happening, and that's very exciting. This is why to be successful, one needs to be a student one's entire life. One must always be learning something new! And right now I am really interested to understand how much of mathematics can be understood through language based on pictures.

Chapter 6
A Conversation with Bernard Sinclair-Desgagné

Can you describe your field briefly?

Within economics, I generally work on the theory of incentives, which can be broadly defined as the formal analysis and design of means to direct the efforts of autonomous agents towards the achievement of given objectives.

Present a paradigmatic example of a model in your field, describing it in terms that are accessible to non-experts.

A key paradigmatic model for the theory of incentives is the *principal–agent model*.

This model was introduced in the 1970s to study corporate governance. Large firms since the late 19th century and the second industrial revolution have been characterised by the fact that investors who provide funding (shareholders, debt holders, banks, etc.) and managers who make use of those funds are generally not the same people. This creates what is called an *agency problem*: some managers may act in their own selfish interest at the expense of the other parties. In the classical principal–agent model, the 'principal' (standing for the firm's shareholders) seeks therefore to monitor and compensate the 'agent' (standing for the firm's CEO) in order to maximise the firm's profit, anticipating that, once a contract is signed, the latter will pursue its best interest.

Since the first formulations, the model has been significantly refined. The principal's and the agent's respective behaviours are now more realistic, thanks to recent advances in behavioural economics. The model can also deal with situations involving one agent and several principals (as when a CEO faces various stakeholders — shareholders, debt holders, regulators, consumers, employees, local communities, etc.), many agents and one principal (as when shareholders must incentivise each individual member of a corporate board), and situations when the agent must allocate his attention and effort among several tasks (fostering quarterly profits while mitigating long-term risks, say).

These extensions, along with substantial empirical and experimental work, have allowed us to gain valuable insights in virtually any context where one party must (at least partly) rely on the efforts of an autonomous second party for its well-being. In addition to corporate governance, settings where such situations are frequent include salesforce and supply-chain management, multinational firms, government bureaucracies, insurance contracting, lending, regulation and law enforcement, healthcare provision, professional and expert consultancy, representative democracy, and even the brain's architecture.

With the help of this example, could you explain why a model is needed? And could you describe what a model is?

Generally speaking, models are simplified representations of objects or situations. Their purpose is to highlight what are considered to be (rightly or not) key contextual features and relationships. As such, they are essential to cope with complexity. As the French writer Paul Valéry once said: "Ce qui est simple est toujours faux; ce qui ne l'est pas est inutilisable." ("What is simple is always wrong; what isn't is unusable.")

Under this rather broad definition, various kinds of representations can be seen as models. The most elementary and concrete ones — in the sense that everyone uses them all the time in daily life — are cases, stories, or what are called 'stylised facts.' Increasingly higher levels of abstraction are reached through pictures, diagrams, and graphs, then formal or mathematical models using formulas and equations. Graphs and mathematical models might involve numbers and computations, or not. The former types are referred to as quantitative models; the latter, and all the other types, define qualitative modelling.

Of course, it is not always necessary, nor desirable, to resort to abstract models. In our daily lives, short stories, images, and case-based reasoning are used quite often enough.

On certain subjects, however, abstraction will make you see things more clearly and think through the matter and your own intuitions while avoiding logical mistakes and communicating more effectively. Let me now illustrate this assertion, in a principal–agent framework.

The novel *Germinal*, by Émile Zola, provides a nice story to begin with.

This novel centers on a labour conflict between coalminers (to be seen here as the 'agents') and their employer (the 'principal'), in 19th-century Northern France. The miners' job comprises two tasks: coal extraction, of course, and also timbering to support the walls and roofs of newly-dug tunnels. Working conditions at the time were dreadful. But at the heart of the conflict is the incentive system. Miners get a piece rate on each tub of coal extracted, while timbering is pushed for through inspections and fines. Incentives to produce are such that miners devote most of their time and energy to this task at the expense of timbering. Destructive tunnel collapses are thus frequent; they inflict significant costs on the company which must then take care of the injured workers and their families. In order to motivate the miners to devote greater attention to timbering, the company decides (not unreasonably) to decrease the piece rate on coal extraction and to pay for the timber. A bloody strike ensues.

Building on this brief account, the following figure — a graphical model — conveys what seems to be the crux of the matter.

74 Dialogues Around Models and Uncertainty

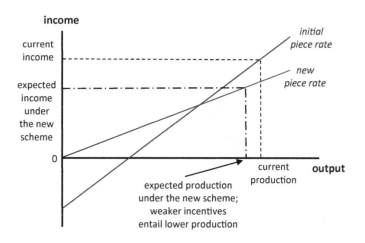

The figure depicts two straight lines — the miners' revenue curves — that relate total income to the amount of coal extracted. In accordance with the story, the initial situation is one where miners are subject to strong incentives to produce (the piece rate on each tub of coal makes for a steeper curve) and weak incentives for timbering (miners actually have to buy the necessary materials, which explains the negative intercept of their initial revenue curve). The company then deems it necessary to rebalance incentives, so the miners would timber their tunnels more carefully. A natural way to do this is to reward coal extraction less (hence providing a lower piece rate, which entails a flatter revenue curve) and subsidise timbering (which brings the curve's intercept up).

The company asserts that the new scheme will not decrease the workers' income. But miners disagree, and the figure shows they are right. Moreover, the most productive miners are the ones who oppose the company most strongly. The figure explains again why: the more productive a miner is, the more distance there is between the two revenue curves at larger outputs, so the more income s/he will be losing under the new method of payment. This is certainly too much for an already drained workforce that cannot possibly accept lower wages and worse living conditions.

Admittedly, it might have been possible to see this, and what therefore lies at the heart of this *Germinal* drama, without the figure. Judging from the discussion between the parties involved in the novel, and between students I have shown the movie to in the classroom, it seems that the present graphical model makes the matter much more transparent.

In addition to clarifying the issue and pinpointing what went wrong, the above graphical model suggests, furthermore, a sensible remedy: as the next figure shows, the company might have increased its subsidy on timbering so that the better miners would not see their income go down.

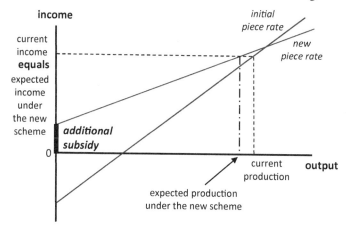

What is the role of mathematics in modelling?

In the novel, coalminers insist they would timber better if they were properly paid for it. In order to precisely establish that appropriate amounts be respectively allocated to excavation and timbering, one would need to translate the above curves representing the miners' revenue into mathematical formulas.

More fundamentally, someone might still challenge this solution because it leads to a decrease in output (while also keeping the cost of labour unchanged, but this is not the main point). Further principal–agent modelling, using, this time, an explicit mathematical formula to represent a miner's preferences, basic calculus, and some elementary notions from mathematical statistics (hence building again a formal or mathematical model), would now come up with the following alternative scheme:

- Keep the initial piece rate on extraction the same, but commit to inspecting a miner's work on timbering if production exceeds a given

fixed threshold (a plausible rationale being that outstanding output might indicate sloppiness on other job dimensions).
- If an inspection then reveals inadequate timbering, a fine will be imposed. But if the timbering is okay, the miner will be granted an extra bonus.
- Fines and bonuses are set so that, *a priori*, a miner wishes for an inspection to take place.

This incentive system differs from the initial one in two aspects: first, inspections that were made at random are now contingent on a miner's output; second, good timbering is rewarded, which was not the case before. Under this scheme, a miner who wants to trigger the inspection of his timbering (expecting, in accordance with the scheme's design, that this will bring him additional income) would seek to produce beyond the threshold; doing this, however, does not make much sense if he does not also seek to do the timbering well. This system thus constitutes an 'ideal' solution, in the sense that it fosters *both* coal extraction *and* better timbering.

This scheme is also quite intuitive. Its relative simplicity makes it suitable for classroom or laboratory experiments. The basic easy-to-explain idea seems implementable in most (numerous) settings in which incentives must be balanced across tasks, such as having managers/executives deliver short-term results while keeping an eye on the firm's long-term business, traders care about both the returns and risks of investments, manufacturers seek to boost sales and enhance consumer loyalty, academics pursue research achievements as well as good teaching, etc. My experience from exposing various audiences to this *Germinal*-based exercise over the past 20 years, though, is that *one would never find out about this scheme without the help of the mathematical model*.

In sum, the fundamental roles of mathematics in modelling I wanted to point out here, and hope to have illustrated well with the *Germinal* example, are: (i) to allow us to precisely grasp and convey the fundamental logic of a situation, (ii) to allow us to make quantitative and qualitative predictions or prescriptions, and, especially, (iii) to allow us to gain new (oftentimes revealing or creative) insights.

What constituents besides a mathematical formalism are part of a model?

A formal model is not a stand-alone mathematical object, like the prime numbers, the torus, or the Fourier series. *Any model is a representation of something, by someone, and for something.*

As a *representation* of some 'reality,' a model must contain elements, and relations between them, that 'correspond' to the entity or phenomenon it is meant to capture. A principal–agent model, for instance, addresses situations involving some party A that must rely on a self-interested party B for the achievement of an objective. In such a model, therefore, one should not simply see functions, mathematical expectations, probability distributions, inequalities, and maximisation operators, but, respectively, preferences over payoffs and effort, production technologies and monitoring devices, outside alternatives, and descriptions of the parties' conduct. This (non-mathematical) correspondence is crucial in order to fix intuition, guide formal calculations and manipulations, interpret and check mathematical results; all in all, to make the model 'speak' to its users.

As a *human construct*, a model also embeds some traits of its creator, such as her knowledge and field of expertise, and her modelling skills and style. These notably show up in the chosen mathematical notation, the level of abstraction, the degree of parsimony (three features that can strongly affect success in mathematical derivations), and, in certain cases, some dose of elegance. Considering the *Germinal* dilemma, for instance, I am sure that a labour economist, an accountant, a firm theorist, a health economist, an operations researcher, an industrial engineer, or a corporate finance scholar, each of them of different ages and academic backgrounds, would respectively write down formal multiple-task principal–agent models that look quite different from each other.

Key features of the modeller's background also show up in the *hypotheses* made and the so-called *black boxes* used. A formal principal–agent model always contains explicit assumptions, or hypotheses, about the agent's and the principal's respective attitudes towards risk, time, monetary and non-monetary outcomes. A finance scholar dealing with CEO compensation will likely put significant weight on monetary compensation, while a health economist examining physicians' remuneration might also insist on professional status and self-respect. Simplification also requires principal–agent modellers to use implicit assumptions and 'black boxes.' Typical ones concern, for instance, the principal's ability to carry out promises (which refers to the presence of external enforcing mechanisms) and the agent's way to exert and experience effort. Someone trained in law and economics might want to spell out the former as soon as she can; a neuro-economist or an ergonomist will show the same concern for the latter.

These remarks do not entail that mathematical models are as prone as other, less abstract models, to personal, intellectual, cultural, or ideological biases. They are not! The main reason lies in the fact that several assumptions are explicitly spelled out in the unambiguous language of mathematics. This makes the resulting analysis more transparent, immune to logical mistakes, and suitable for rigorous challenge and discussion. The only way some biases might persist is when they remain hidden in the 'black boxes.' Model users should thus constantly be aware of, and acknowledge, these black boxes, and nuance their conclusions accordingly.

Finally, a model always reflects some *use or intent*. It is generally possible to tell, based on seeing the mathematical expressions, whether a model is intended for empirical or theoretical research. Computational and theoretical models are constrained in their formulation by mathematical tractability considerations: empirical (econometric or experimental) models by data availability, measurement, and control imperatives. These constraints, together with the ones set by the state of the art in the concerned disciplines (like utility maximisation as the enduring standard representation of rationality in economics), will necessarily stand out in the ultimate form of the model.

All in all, these remarks explain, perhaps, why great mathematicians or statisticians are not always good modellers. There is much more than maths in a mathematical model. Using here a distinction made by Blaise Pascal at the beginning of his *Pensées* (published in the 17th century), I'd say that an effective mathematical modeller must naturally call upon "l'esprit de géométrie" (i.e. the 'mathematical' mind, as it has been translated in English), but somewhat also (and rather increasingly, when one moves towards policymaking) "l'esprit de finesse" (or the 'intuitive' mind).

What is the role of language in modelling?

Mathematical modellers must speak at least three languages: a natural one (English, French, Spanish, Chinese, etc.), of course, plus the technical tongue of their scientific community (economics, finance, accounting, sociology, psychology, operations research, engineering, law, etc.), then the language of some germane mathematical area (real analysis, linear algebra, optimisation, game theory, mathematical statistics, etc.).

The relative contribution of each language varies at the successive stages of modelling.

In the preliminary phase, when it is not even certain what the problem is and whether a formal model is at all needed, the natural language leads the way. At this early point, the goal is to get a finer grasp of the issue, formulate a clear (often shared) research question, and lay an intuitive background. The latter is best expressed by a simple story, like the *Germinal* one, or a crude but sensible empirical proposition (which economists call a *stylised fact*).

As I said above, a short case or a stylised fact is an already simplified account of reality, hence the formulation of some sort of model, which often points in the direction of a more abstract representation, is the next step. As one takes on this next stage, the technical and mathematical languages take over. Both languages bring clarity and precision. On the one hand, mathematical tractability imposes further simplifications that must make sense from the scientific and intuitive viewpoints, while not making the model entirely tautological. On the other hand, mathematical abstraction often reveals surprising similarities with other contexts and problems.

When the model is next manipulated, it will first 'speak' the language of mathematics. As the history of mathematics teaches, well-chosen notation is key (try to multiply and divide using Roman instead of Arabic symbols, for instance) and will make the model do so more easily. In a principal–agent analysis, one normally has to check optimality conditions for, respectively, the agent and the principal. Although economic intuition can help here, these conditions can only be rigorously established using mathematics.

Finally, one must interpret/translate the obtained mathematical expressions, hoping they will deliver valuable insights. This is the exclusive domain of technical and natural languages. In my opinion, being finally able to express formal conclusions intuitively in a natural language constitutes a key step to ascertain the model's reliability and relevance.

Models are often said to represent a target system (typically a selected part or aspect of the world). Does this characterisation describe what happens in your field? If so, could you say how a model represents its target? In other words, how do you understand the model–world interface?

Again, a model is always a simplified representation of a given aspect of the world that is deemed of interest. This being said, I would like here to make a remark, followed by an important caveat.

In economics, one distinguishes between descriptive and normative models (though the line between the two is sometimes fuzzy). Models of the former type — the largely prevalent ones in the natural sciences — are supposed to explain and/or predict some specific facts; the model–world interface is ruled by the possibility to observe what the model precisely says about the world as it is. On the other hand, normative models (used notably in clinical medicine, engineering, and economics) are meant to deliver acceptable and implementable schemes; such models must then also include concrete means to act upon the world *and* some criteria for deploying those means. The principal–agent model alluded to above to deal with the *Germinal* dilemma was of the latter type: the criterion was the company's profit, constrained by the workers' discretion in choosing their effort allocation, and the means were the timbering inspections and the coalminers' compensation.

Since normative models are in essence less subject to the verdict of data (they are, of course, to the requirements of logic and, although sometimes not enough, to the findings of science and the lessons of experience), a warning I would again make is that users of such models must always beware of their underlying framework, implicit assumptions, and black boxes. This can be difficult, especially when these assumptions are routinely pushed through without ever being questioned and spelled out. Tracing back the model's history and original purpose, confronting with new stories or stylised facts, might then be useful. Ill-conceived executive compensation, for instance, is often blamed on the recommendations drawn from an early principal–agent model. This model focused on the executive's total effort while ignoring the distribution of effort across tasks. Yet, plenty of (real or fictional) stories, including the *Germinal* example, could have shown the pitfalls of this simplification.

What is the relationship between a model and a theory?

My view is that, in economics at least, a theory is a set of propositions (empirically testable or normative) on a given subject (contracts, regulation, managerial incentives, justice, and so on) which are supported by models. A theory thus appears and develops simultaneously with the production of new models and the refinement of existing ones.

A useful analogy could be drawn here between a theory and its models, on the one hand, and a literary genre or painting school and their associated writings or paintings, on the other hand. Impressionism, for instance, refers to paintings by Monet, Renoir, Sisley, and many others. Thriller as a genre of literature is defined by a large set of individual novels. Similarly, the theory of incentives contains many models: principal–agent models of one or many agents facing one or many principals, of an agent attending to several tasks, as well as models of the effect of information provision on behaviour, models of the impact of social status, models of peer pressure, models of leadership, models of social networks, models involving norms, standards, intrinsic motives, emotions, etc. Each model supports propositions which altogether constitute the theory.

This analogy suggests that *pertaining to a particular theory puts additional requirements on models.*

First, in a given theory, models have to conform to some codes or standards. To be called impressionist, a painting has to exhibit a peculiar style and some specific techniques. All thriller novels must, in different ways, expose the reader to mystery, sustained tension, and surprises. Similarly, in the theory of incentives, all models agree with the paradigm of economics, which posits 'rational' players who seek to improve their individual well-being.

Secondly, within a theory, models must support propositions which are not incompatible with each other. When inconsistencies appear, some model has to be discarded from the theory, or the theory itself has to be amended or even abandoned.

What is the aim/use of the model: e.g. learning/exploration, optimisation/exploitation?

Descriptive, predictive, and computational models aim to explore and make us learn something. Normative models also contribute to knowledge, through the good and bad experiences their implemented recommendations generate. But their main purpose is to optimise a criterion or exploit an opportunity.

In case you use computer simulations, what is the relationship between simulations and the model?

Like the graphical model used above, simulation — thanks to the concrete numerical results it generates — can act as a preliminary step towards building a general mathematical model.

Simulation can also be the only alternative possible to make a model speak when mathematical difficulties preclude finding a clear-cut analytical solution.

In sum, simulations can be used as a helpful device in model building, or as a computational tool to draw insights from an existing model.

What is a good model?

Quoting my former colleague Arnoud de Meyer, I would say that a good model — like any good research, actually — should be *relevant* (dealing with an issue of interest), *rigorous* (given the current standards of the involved scientific community), and, most of all, *revealing* (i.e. bringing us beyond what is obvious).

The latter request reminds me of my answer to a student who once asked me the same question — "Sir, what is a good model?" — just after I had run the *Germinal* exercise. I ended up saying that a good model is a formal representation yielding insights we might not have gotten without it; after we found something, however, we should be able to explain what it is without referring anymore to the model. (I found out afterwards that this view dates back to Alfred Marshall, who made a similar statement a century ago.) In this sense, a good model is a device that not only builds on intuition but also pulls intuition forward.

When considering two models of the same issue that deliver seemingly equivalent results, finally, a scientist will generally prefer the simpler (this is the so-called 'Occam's razor criterion') and more elegant one. Simplicity refers to the parsimony of assumptions (which often allows further extensions) and the mathematical sophistication involved. The elegance of the mathematical notation and the way results are derived matter as well.

How would you define risk? And how does the model help us understand risk?

The theory of incentives is rooted in decision and game theory (while now drawing increasingly from psychology, sociology, and related disciplines). It therefore defines risk as a range of possible outcomes (good or bad), respectively, weighted by probabilities.

In a principal–agent model, the principal is the initial risk bearer. The probability distribution of outcomes, however, is influenced by the agent's non-observable effort. (In some models, the agent might also select the set of distributions to be impacted by his effort.) The setting is thus one where risk is not an 'Act of God' or something totally unrelated to human actions, like the occurrence of a storm, the results of national lottery, or the value of a stock market index (from a small investor's perspective). Risk depends here crucially on what the agent does. It is 'endogenous' rather than 'exogenous.' And if the agent is not properly motivated, he might act carelessly from the principal's viewpoint.

This situation is familiar notably to insurers, who refer to it as *moral hazard*: someone who is overly sheltered from risk will tend to somewhat ignore it, which can be detrimental to himself and other parties. Moral hazard can indeed be seen in all spheres of society, from inattentive drivers to people eating unhealthy food, to traders inflicting unwanted volatility on investors, to banks being lax on credit (knowing they are so big, a government will never let them go bankrupt), to nations running deep into debt (with the belief that they will be rescued if things go bad). The main contribution of principal–agent models is to analyse how a principal actually deals with, or can alleviate, this risk factor, thereby helping us understand how this risk is, or can be, managed.

Typically, the principal will want to transfer some of her risk upon the agent, in order to bring the latter's decisions into line with her own objectives. The amount transferred is limited, though, by the agent's own capability to bear risk and his ensuing demand to be compensated for this. There is, therefore, a fundamental trade-off between risk-sharing and incentives. Analysing this trade-off lies at the heart of principal–agent modelling.

How would you define uncertainty? And how does the model help us understand uncertainty?

In decision theory, uncertainty refers to the impossibility of representing a venture using definite probabilities assigned to pre-specified outcomes. Principal–agent models can deal with this in many ways.

When outcomes cannot all be listed *a priori*, contracts between principal and agent will usually be vague on what to do in case an *unforeseen contingency* shows up. This calls for allowing the parties to then renegotiate their initial agreement. There is now a wealth of principal–agent models of such 'incomplete' contracts subject to potential renegotiation.

Thanks to recent advances in microeconomics and decision theory, moreover, some recent principal–agent models also confront the parties with what is called *ambiguity*, i.e. the simultaneous presence of several probability distributions (whatever the agent's chosen effort); the agent (and sometimes the principal as well) is then endowed with some degree of ambiguity aversion.

Finally, and perhaps more importantly, it turns out that the compensation packages which hold in practice are often simpler than the ones predicted or prescribed by the theory. This suggests that our models might well deliver schemes which are 'optimal,' but the actual principals facing a complex reality and uncertainty rather seek devices which are *robust*. Over at least three decades, another active stream of research has thus been focusing on the robustness of incentive contracts.

What do you consider to be the work/result that has had the most significant impact on your field? And why?

The most influential paper in the theory of incentives is perhaps Jensen and Meckling's: *Theory of the Firm: Managerial Behavior, Agency Costs and Ownership Structure*.[1] Published in 1976, this article has been quoted 52,200 times according to Google Scholar. It has notably been used as a basis for establishing executive compensation.

These were the early days of principal–agent modelling. Unfortunately, several people — usually very critical (with good reasons) of current practices in executive compensation — tend to reduce the theory of incentives to Jensen and Meckling's article. But the theory has of course evolved and become much richer over the past 40 years, thanks to numerous multidisciplinary challenges, influences, and insights (coming from accounting, psychology, human resource management, corporate law, etc.), improved modelling techniques (drawn notably from game theory and statistics), advances in empirical research (the advent of experimental economics), and so on.

One subsequent contribution which has had a significant impact on the field and my own thinking (Google Scholar reports about 5500 citations) is Bengt Holmstrom and Paul Milgrom's first model of multitasking.[2] The *Germinal* exercise proposed above aimed indeed to add to the research agenda set by these authors.[3]

In concluding their milestone paper, Holmstrom and Milgrom make the following statement: "Our approach emphasises that incentive problems must be analysed in totality; one cannot make correct inferences about the proper incentives for an activity by studying the attributes of

---

[1] Michael C. Jensen and William H. Meckling (1976), "Theory of the firm: Managerial behavior, agency costs and ownership structure," *Journal of Financial Economics*, vol. 3, pp. 305–360.

[2] Bengt Holmstrom and Paul Milgrom (1991), "Multitask principal–agent analyses: incentive contracts, asset ownership, and job design," *Journal of Law, Economics and Organisation*, vol. 7 [Special Issue: Papers from the Conference on the New Science of Organisation], pp. 24–52.

[3] See Bernard Sinclair-Desgagné (1999), "How to restore higher-powered incentives in multi-task agencies," *Journal of Law, Economics, and Organisation*, vol. 15, pp. 418–433.

that activity alone. Moreover, the range of instruments that can be used to control an agent's performance in one activity is much wider than just deciding how to pay for performance." More than ever, this now sounds as a call for modesty for principal–agent modellers, and as a prescient warning against relying on a single model.[4]

---

[4] The latter point is further developed and conveyed persuasively in a recent book by Dani Rodrik, *Economics Rules — Why economics works, when it fails, and how to tell the difference*, Oxford University Press, 2015.

Chapter 7

A Conversation with Sir Brian Hoskins

Can you describe your field briefly?

I'm a mathematician originally, and I came into using fluid mechanics essentially to understand how the atmosphere works. So, I've been involved in what's often called dynamic meteorology, that's how the atmosphere moves, but essentially in weather and in climate.

Present a paradigmatic example of a model in your field, describing it in terms that are accessible to non-experts.

I looked at this on the way here and thought that I would give you two, if that's alright?

That's more than alright.

One example is given by what I've worked in traditionally, and the other is a slightly more general thing in my subject area that I interact with a lot and have used. So, the first one is a model; all the models, I think, in my subject area, are based on Newton's Laws of motion, equations of thermodynamics, hence they're based on scientific laws, and then encapsulating those for a particular, theoretical idealisation of, say, a weather system in middle latitudes. So that's type 1, which to me is actually an idealised model of a middle latitude weather system, getting it down to essentials, so that's reducing it down. Whereas the other type of model that I've been very much interacting with is one which is trying to use those laws to represent the evolving weather, maybe in terms of even from shorter timescales through to climate, so it's representing the atmosphere in terms of those basic laws, but then representations of physical processes such as clouds, biological processes as necessary on the surface, maybe the chemistry of the atmosphere, and then interacting with the ocean system as well where similar…

Would you say that the key differences between these two types of models are on the timescale and geographical scale, or…?

It's also on the motivation: one is as an applied mathematician getting an idealisation of a weather system, whereas the other is putting in quite a lot of complexity in order to simulate better.

That's interesting. We will come back to the aim of the modelling later. So, with the help of these two examples, could you explain why a model is needed?

Why is a model needed? Well, one is as an academic thing, to try and understand how the atmosphere works; but then I think in my subject area, it's always very much feeding in to the practical thing of actually producing a better forecast of weather or climate; so it then feeds in to what are the essential ingredients in a more complex model, and also a framework of diagnosing what comes out of a more complex model, or what comes out of atmospheric data, for instance. So, one is an idealisation to help the whole theoretical basis of the subject if you like, and the other is its diagnosis.

Now, we will try to understand a bit better 'what is a model,' with a number of questions. So, the first question is what would be the role of mathematics in modelling? Or quantitative methods?

In the first type of model, one is actually putting into a mathematical context Newton's Laws of motion, the equations of thermodynamics. But then the mathematics is, particularly in this idealisation, and in terms of approximating those equations and producing something which is consistent with the phenomena, used to communicate the scientific laws in the context of the system of interest, so producing an approximation, down to a level which is actually soluble, either as a simple computational problem or maybe even analytically.

So, you would say that the core of the model will be mathematical because it will be equations?

That's right. So that is in set one, whereas in the second one — which is why I thought I would produce the two — the whole numerical approximation to those equations and the inclusion of the representation of those physical processes, etc., that whole thing becomes the model. So, in one case, the mathematical set of equations is the idealisation, is the model, and in the other, the rather more complex thing including its discretisation and the representation of the other processes becomes the model.

So, you would say, somehow, that in the second type of the model, which seems to be, at least to me, much more complex,...

Yes, much more complex.

Because it's so complex, it cannot be formulated simply in mathematical terms, or in equations, and then you will use simulations?

It is actually described in mathematical terms, the discretisation is clearly a mathematical thing, and the representation of all those processes would then be given a mathematical form. So, it is actually represented in mathematical terms in order to actually put it on a computer. But each of its parts is in a mathematical form. But it will include parts that don't just come directly from scientific laws — some of them come from observations, I mean if you're representing a cloud, you're not coming from a mathematical representation of every detail of that cloud and then simplifying it, you are producing an idealised model if you like of a cloud and then using observations to get parameters for that model.

Would you say, then, that this second type of model is a sort of meta-model?

Yes, I think it is probably that. It would be referred to as a general circulation model or climate model or a weather forecast model.

Do you have models for each little entity?

Yes, that's right.

And then they are put together?

That's right, usually in the context of representation of the basic physical laws themselves, but then little sub-models for various other things that are added on.

So, for instance, you would have one for the clouds and one for the ocean…?

Yes, that's right. Now, the chemistry, if that's included (personally I haven't done that myself, but I interact with those who do), that would then come in terms of chemicals that would be advected with the velocities that you have in the model, but then you would have a little sub-model of how those chemicals react with one another.

So, the link between all these models in this meta-model is governed by physical laws?

It is governed by the input to those physical law equations.
 I hope that's not too complicated, but I thought I would produce the two that I've been associated with.

No, I think it's really interesting to have this dual approach. So, besides this mathematical formalism, you mentioned physical laws. What would you say are the other constituents of the model?

The other constituents of the first type of model are purely then approximations and then mathematical techniques like looking for normal solutions, etc.

As regards approximations, do they include what in other fields is referred to as calibration?

No, it's different from that because we live on a rapidly rotating planet, then there are certain balances that occur between, say, the winds and the temperatures. So, you can either take a set of equations that doesn't include those, sometimes those are called the primitive equations although they're not that primitive as there are still approximations in them, but there are equations that build in that balance. So, there are certain balances such as…one that is hydrostatic balance in the vertical. Most equations that I would use include that as already in there, so the hydrostatic approximation is made, which means you throw away the vertical acceleration of the atmosphere, you don't include that directly. But then, on a rapidly rotating planet, there are approximations associated with the balance between winds; so, the wind goes around a low pressure system, in the first order, it goes around, so you can say well the first approximation for wind is it goes around, but then you go to a higher order approximation that it actually doesn't quite go around, but it's an approximation based on a small number that is associated with the rapid rotation of the planet.

So, what you refer to as an approximation, can it be seen as an assumption or simplification?

Yes, it can be seen as a succession, so you have a small number, and then you do an expansion of that small number and you retain certain terms. So, that's the first step, which very much needs a mathematician. Now, the second set of models doesn't include that sort of thing in general.

I'm making life complicated having the two sets, aren't I?

What is the role of observation or data here?

In the first set, observations enter very much into framing the model. So, you say there are these phenomena that occur, they occur on these sorts of space and time scales, in these sorts of latitudes, and that sets the framework then. And what are these sorts of properties of those systems which will enable them to guide or which will guide your approximation? So, in the first set, the observations guide the framing of the model and the approximations made. In the second set then, observations underlie what you're going to represent explicitly and what you're going to 'parameterise.' So, there are those processes represented explicitly, and those will derive from the basic laws, and there are those things which will be implicit. So, in a model like that, you would carry water vapour as one of your variables, but then the actual processes in which the condensation, or the evaporation, of water occurs would be parameterised, and those parameterisations then very much involve observations of how this actually occurs and are simply mathematical models with parameters based on observations.

So, would you say that in the second type of model, data are really part of the input?

So, observations guide the inputs...

I mean the inputs being the outputs of the little models?

Yes, so, let's take a little model of a cloud, for instance, if you're going to say there is convection, there is a convective cloud, a bubbling up cloud, then that would involve a single column in the model, and you say well that plume as it comes up will entrain certain air from the surroundings and there'll be a parameter that actually talks about that entraining and the actual instability in the vertical will actually involve, having been guided by observations, how that goes in and then the plume will get to a certain level and then it will detrain and a certain amount of the moisture will be taken out and a certain amount condensed, so those sorts of things will be determined by looking at the observations and help that guide your little sub-model.

This makes sense. Does language have a role?

I saw that question and I quite wondered what you mean by that.

It might be calling certain things by a certain name, does it have an impact on the way you model things?

I don't think so. I was trying to see how this might work and I can't see language playing a role.

Or qualitative aspects more generally? Or is it really a quantitative subject?

Well it's a quantitative subject, but I think there is...well, there's no doubt that in framing the first sort of model, one is getting the essence of something, which involves looking at it, maybe some sort of language...this is the important aspect; in the second type, I think language can mislead you in what you're doing from the output of the model — you might think that you're doing more than you are.

Does this relate to the output and the communication of the output?

Ah, well, very much so. What do you think you're actually producing from the model? It comes to what can you trust in that model, as in the output of the model. What framework do you have for actually putting together the results of that model, and how then do you communicate it? Now, if you take the weather, the second type of model, then one simple way of communicating is via a front in the atmosphere — you know you see those on a weather map. Now, that's an idealised model that we know doesn't actually work, it doesn't look like that. It actually came from the First World War where they had fronts and they thought, well, this is the sort of thing that's going on in the atmosphere — that there's a frontline between them that moves a bit. So, there's an idealisation there, which is given a word, and that sometimes can almost take over people in that they try and fit what comes out of the model or what they see in the atmosphere too much into that model. A model can be very useful for communicating, but it can actually hold you back — such a word like that and the concept that goes with it can almost hold you back from actually seeing more to it than you've actually got going on in the atmosphere that you could understand or coming from the model than your simple word 'front' allows you to put together or maybe even communicate.

So, you would say that language could make the researcher a bit myopic on the outcomes?

It can...up to a point it can help, and sometimes it can then hinder if it's actually taken too strictly. And I remember an observational programme — they were trying to fit, going up in aircraft to look at fronts in the atmosphere, and I was appalled that they were actually trying to fit their observations into a model that was the previous model and was too simple for what they were actually observing — they should have been taking things beyond (but I know you're not so concerned with those observations).

What about notation? Do they play an important role in the modelling?

I don't think that notation plays a crucial role in either of my modelling categories.

Models are often said to represent a target system. So, does this characterisation describe what happens in your field?

I think, in the first case then, you really have got a target system and you're trying to idealise that, what the model is focused on, and that sort of system.

Could you just say the question again, so I can think regarding the second type?

Does the characterisation that the model represents a target system describe what happens in your field, and if so, how does it represent the target?

So, I think the first model is to represent a target, a weather system. The second type, which is more, you might be running a model to say 'how do sea surface temperature anomalies now affect next summer's weather?', for instance. So, that might be the focus of that sort of model. There is a question there, and that question guides the actual framework in which you would use the model. So, you typically might use a model where you have a control without that anomaly, and you run that control many times, and then you run an anomaly case and you run that many times, and you see then whether there is a difference between this set and that set? And then you could say that difference is due to the sea surface temperature number we put in. So, the models are used in this sort of way, for a target I suppose. Sometimes, beyond that, similar models will be used just for saying what's our best forecast for the seasonal weather? Or it might be a question of, with increased greenhouse gases and carbon dioxide in the atmosphere, if we take one particular aspect of the climate system, how might that change? And again you might use a similar formalism, a structure for attacking a similar problem.

So, would you say that in the first type of model, the model of the atmosphere, you would have the model as an understanding, a description, and then really you have the target and then you try to represent this target as best as you can, whereas in the second type of model, the objective is more to do with what is going to happen under certain scenarios?

Yes, I mean the first sort is looking at a particular phenomenon, with a particular structure to it, and one is using that model not to get as close to reality as you can. You're using it to get the essence of that thing, and then you might add some, then say, well, we've got the essence, let's suppose we can look at the sensitivity to certain parameters, and we can say, well, this model might be a dry run of a middle latitude weather system, could we actually build on that model and say what would happen, well we know there is latent heat release, there is rainfall, what might that do? But again, not giving it the full complexity, can we use that and build a little to see what the effect might be? So, that's building up on that. In the other model, one tends to be wanting it to be close to what you observe, so the essence of it is to try and make it complex enough — people often talk about "sophisticated," which means making it overcomplex really, you'll hear people in my area talking about a sophisticated climate model, which sounds as if we've made it complex enough that no one can understand it, so they're bound to believe it. But you do try to add, if it is deficient in some way that is important to you, you actually try and put something else, which you think might be the reason for that deficiency, to make it more close to the simulated system.

Is it because in the second type of model, we don't know what can have an impact, and therefore, we try to capture everything we can at the beginning?

Well, I think, with the second type, you do say well, we think these are the important ingredients and we include those, but then, we find the result is deficient compared with observation, so we then say, well, is there something we're already representing which we're not representing well enough?

And then you go back...?

That's right and you might say, well, we think the convection, the convective process I described before, is not being described well enough, and we could say, well, let's look at observations again and have a particular observational campaign to try and see how the context of a certain region of the earth, or a certain typical weather system, how these processes are actually working, and maybe we aim to improve our representation, or you could say, from looking at the real world, if we have a summer like 2003, it's actually very important that the vegetation is actually tending to wilt as that season goes on, so you don't get the same water coming out of that, so maybe we actually need to put dynamic vegetation into our model and that that's the crucial thing. So, you actually put a new whole process you're going to model as a sub-model and you put it into that. The trouble is, with that sort of model, you do have this thing of the structural nature of the model, and is there structural sensitivity if you add a new component — has that changed the whole structure of the model or maybe some bit we don't have that would actually change the whole structure of the model? So, by really investigating the model without that sub-component, we're actually going in a space that is not very relevant, we should actually have that extra process.

Maybe here, it's good to talk about sensitivity. I guess in the first type of model, it is closer to a sort of robustness analysis where you tweak some parameters and then you observe what is happening, whereas in the second type of model, it seems that the sort of sensitivity analysis is sensitivity towards taking account of a block of information. Am I correct?

Ok, so in the first type, the sensitivity analysis, with the first bit, as you say, if you have this simple model and what is the behaviour as a function of wavelength if you're looking at structures that move and grow, is there a preferred wavelength where you can look at the sensitivity then, there are basic parameters like the basic stability of the atmosphere — how does that change this weather system if you change that stability? The rotation rate of the earth — how does that change it? And the strength of the westerly winds — how does that change it? So, you can look at the basic parameters of your model, the background parameters of your model, and say how sensitive is it to that, and that's interesting information. Then you could look at, and perhaps get an indication of the sensitivity to things not included, such as latent heat release. Can we get some idea of how sensitive this might be to the inclusion of something else in a very simple way? So, we can use that. Then in the other model, sensitivity comes in a couple of ways I think: there is actually comparison with your target system, for which of course we only have maybe 50 or 100 years of record, and then we have a system that has variability on all timescales, and so it is quite a problem to actually, say, compare your realisation with the real system. If it's a weather thing, then you can say well, there is sensitivity to certain parameters, so we might investigate those, represent those sub-models; there's also sensitivity to initial conditions if you're doing a weather forecast, and we want to explore the sensitivity to initial conditions.

The initial condition being the calibration?

No, the state..., so the sort of model I'm looking at here, you have this model, but then you say well, they are solved on a time-stamping, so you have initial conditions.

So, for instance, for the little cloud models, the initial conditions would be information about...?

Well, usually those cloud models just take the state of the atmosphere that you give them at any time and then they work with that. So, there's a sensitivity to parameters in those models, but then there's the sensitivity of the whole model to the initial conditions, which, if it's just a weather forecast model, will be the initial conditions in the atmosphere — so, you have stations around the world that observe winds, temperature, and everything, you have satellites, and all those are processed together to give you an initial state of the system. And we know that...in fact, I'll come back to the initial state, I'm not sure if that's relevant. But then we're aware that that time-stamping will…and what you get for the weather here will be sensitive to the initial conditions. So then, you can try and handle that sensitivity to initial conditions, and that is done by getting a range of initial conditions which are consistent with the observations, and then running the forecast model many times, so typically running a forecast model, maybe 52 times, is done for every 10-day forecast. So, it's run 52 times, trying to span the space of initial conditions, getting in some sense, the biggest departure from the initial state; so there is sensitivity to initial conditions and then there's sensitivity if you are running a weather forecast model, then if it's more a climate model, there would be the oceanic state as well and the state of every other part of the system that you've got and how that would depend. So, there's sensitivity to parameters, and there's sensitivity to initial conditions, and what is done is to try and do an ensemble which tries to explore all those, you can't really, but you try and span the space as best as you can and get some idea of the sensitivity. But that's if you're running an actual simulation. If you're trying to understand the model and improve it, you might try and say, well, where is the biggest sensitivity to the parameters we already have? Which one is my representation of climate, say, most sensitive to? And try and get some idea of where the biggest sensitivities are.

What about the various parameters towards which you could be sensitive? Are they independent, or if you tweak one, do you change everything?

No, that's a basic flaw, I mean, the pretence often is, if you have a multi-dimensional space of parameters, then how do you tweak one and see how much they are dependent on one another? So, usually, there is an attempt to try and tweak them independently, with some indication of whether they are really independent, but…

Because somehow in your mini-models, they are all connected and so…

They might well be, and the optimum, with this set of other parameters, then for this might be this, but actually it's one of these that's wrong and this one was right in the first place. That makes it extremely difficult.

I think you have similar processes occurring in various places around the world to develop models, and if one of them goes into a blind alley, then maybe you hope others haven't. There's a black art here if you like.

I think we can slightly expand the next question. What is the relation between model and theory? Also, because you mentioned earlier scientific laws, how would you put in perspective model, theory, and scientific laws?

We have the basic laws which then, with various approximations, can be used in our models. I think I take the theory as the thing we build up using my example 1-type models, those give us a basic theory. They may be simplifications, they will be based on simplifications of the basic laws, thinking about the sort of things we're trying to model; the idealised models of certain parts of the system I think then give you these models like these, and the ensemble of those becomes the theory.

And this would be a sort of common feature of models for the same thing?

Yes. And as the theory might have certain...it might be a middle latitude weather system and you have a body of theory from that; but then there might be a body of theory that applies to tropical cyclones; there might be a body of theory that applies to the impact of additional greenhouse gases in the atmosphere which might start from very simple radiative transfer type things, and so you have the representation of the basic physics in terms of modelling how the solar radiation would come through the atmosphere and how the long-wave radiation would go the other way. I mainly come from the field of fluid dynamics, but there's other stuff. So, all those would tend to give you a body of theory which has different parts of the subject that it's actually addressing.

And the theory would be studied through different models?

Through different models… I think I like to fit it together with models and observations, so the observations are always there, so you have — one of the things which I don't think is done enough, I gave a lecture on this in 1983 and I'm still talking about it — is a model hierarchy. There are too many observations, there's theory with rather simple models, and there are the very complex models, and they all sit apart. To me, a hierarchy of models is almost, in some sense, the theory is there as well, the hierarchy of models almost encapsulates the theory we have, too. And the complex models are really just this end of that hierarchy, and they do always take account of the observations. So, we have our body of theory and our hierarchy of models and the observations, and those really are melded together in some sense in order to make progress. When you're diagnosing the complex models, you need a simpler model framework, which if you like is theory, and I view those together, in some sense, as a way of diagnosing that model, and if you want to improve the model, you have to go back to that understanding really to know how to improve the model. One example I have is that all the best ingredients were put together for a new climate model, and it didn't rain over India — the summer monsoon didn't occur. Now, the way those involved in the complex modelling tend to go, let's try and change some parameters over India, the land surface is wrong or something, but to me, if we go back to this theory, we can see actually where does this rain come from? Well, it's air that crosses the equator. How does it cross the equator? And there are certain ideas of that because it's actually not very easy for air to cross the equator, and then it doesn't return to the southern hemisphere, it stays in the northern hemisphere. Maybe actually what you got wrong is something over the southern Indian ocean. So, theory helps guide you on how you can improve your model.

So, going back to models, we mentioned this a little bit earlier on, what is the aim or use of the models? I gave a few examples, such as learning, exploration, optimisation, exploitation, etc.?

I think probably it has two uses: one is to enhance this body of theory if you like, and the other is to actually simulate the weather, or the seasons, seasonal weather or climate, in order to interact those with society and its uses. So, it's the two ends, maybe it's a bit of a spectrum, but it is really the building up of the theory or else producing predictions or projections which are of practical use.

But, from your diagram, it's very striking that the first type of models, because they help understand the theory and build on the theory, have a direct impact on the second type of models.

Absolutely. And the second also have an impact on the first because it says we need to understand this more, or this is behaving in this way, so that really puts some impetus on this theory to actually understand those more. So, this one helps represent, if you want to reduce those billions of numbers down to something that you can both understand and even give to someone else, one has to use this body of theory to do that.

This is really about research and why the government should fund academic research?

Yes, absolutely. You may see this perhaps as a diversion for you, but to a certain extent, those policymakers and those who fund research would like to think that we understand climate and climate change, now this global problem, and what we need is what's going to be in my back garden in London in 2080. So, you know, forget this global thing, we've done that. Now, just make your models very refined and we can resolve my back garden and say, well, at 12 o'clock on the 1st of January in 2081, so let's get the real impact of climate change. So, you have to tell them, well I'm afraid that model is going to be no better than the information that's provided on its boundary conditions, and in the end, you need the whole globe, and if you've got the thing wrong there, I have various rude ways of saying this, but if you have large-scale rubbish, you can get very detailed rubbish. So, it can be misleading. Understanding the whole problem is necessary before information can be given, and actually it's important to tell somebody who's trying to use this, don't trust this information as a real prediction, it is the sort of thing you can have, not a prediction of what you will have. The whole framework there is needed for that.

We'll come back to that when we speak about uncertainty I guess. You mentioned several times simulations and computer simulations. So, what is the relationship between simulations and models? It seems quite intricate.

Yes. Well, the way it's used in our subject is that models, these complex models, are used for simulations. And then, we have this hierarchy of models which are used more for the basic understanding of how you use the complex model. So, simulation is done with these high-end models that are close to being used for practical purposes.

In other fields, in cosmology apparently but also in clinical medicine, the development of tools, like telescopes or computers to look at DNA, etc., has had a massive impact on the modelling and what we can do. It seems that it's also the case elsewhere.

Yes, it's all happened over my research lifetime basically. There have been two major things: one is the ability to observe the atmosphere from space. That's been huge because suddenly one had this global domain that is available and a huge amount of information coming down. Second is the ability to represent the discretised system in so much more detail than was possible, and it's happened at a breathtaking speed. In some ways, I think the computational ability has almost fed the ability to do the complex model perhaps faster than our ability to really know enough to use those models.

Do you think it's very positive…?

Oh, incredibly positive, absolutely. And we're always being torn, at this more complex end, between aspects such as do we put finer and finer resolution? Do we put more and more processes (i.e. include in the model more chemistry, biology)? Do we run the models for much longer, so we're really doing paleo-climate simulations? Or do we run many more members of an ensemble to look at the sensitivity?

Could you define ensemble for us, as this is a very specific term not used in many fields?

Ensemble? That's many runs of the model with maybe different initial conditions or different parameters. That's the ensemble approach, which is very widely used now. So, the increase in computational power has been amazing, but to a certain extent, I think, it almost became that one was able to do this and the thread leading from the basic theory to there wasn't quite as strong as it should be; actually, much of my career I have been trying to use models that could be used towards this end but actually using them for experimentation to increase the theoretical basis. So, in the idealised model of a middle latitude weather system that I mentioned, well one of the things we did was use this increased computer power with models that are fairly idealised in some sense, but to actually extend the theory and so, not pull it in full but only partial complexity.

So, you mean that ideally you have the model, then you use the simulation, and then from this, you should go back…so like a back and forth?

Well, maybe not a back and forth, but, ok so this simple model of a weather system is on a plane, and then you can say well what happens if we put it on a true spherical domain rather than on a plane? How does that change it? You can't do it analytically, but you can do it using a computer. But you can set the problem up very much guided by the analytical thing on a plane, so what is the impact of spherical geometry? What is the impact of getting rid of those assumptions about the balance between the wind and the temperature? So, you can do that. What happens if we actually allow for finer resolution which resolves features like fronts that weren't resolved if you just discretised on a broader thing? Then you can say, well, how important is the structure of the westerly wind that you have a jet, that it's actually not a uniform flow? So, on this domain, you can then, using computer power, do that naturally. So, that's the sort of way we've built it up. And then you can say, well, if we allow some latent heat release, allow some rainfall to occur, how does that change the system? So, you can build it up; it's still not simulating something you observe, it's actually building up the theory.

What is a good model?

What is a *good* model? Some of these theoretical models are good, in the sense that you've said you want to do an idealisation and that is a good model with the idealisation that you've said you wanted to do, in that it seems to encapsulate the basic processes of this more complex thing, and it seems to have a maximum simplification that still allows that.

I would think that's a good model. And if it's a good model, well the model itself…then if you, on this sort of hierarchy of models, I think this is a good model if it's well based on the previous ones, and you are well in control of how to experiment with it, then I think that could be classified as a good model. But defining anything up here as a good model is dangerous because it might appear to simulate something well, but it might simulate other things poorly, and then what do you mean by simulate well becomes the point, because you often see this, people say, well, my climate model, if I look at the 20th century, it's simulating it well enough that we can use it for the 21st century. Well, what does that mean? What is 'well enough,' and what basis would you have for saying so?

That's really interesting. So, would you say that you have a good model for some purpose, and the same good models might not be good for other purposes, depending on what you want to do with the model?

I think there's a danger in the word 'good.' If I move beyond my idealised world, the word 'good' is dangerous in that it suggests something that those who don't know the details of what's in it might trust it more than you really should.

It's a dangerous word. I'm very suspicious of models.

Well, a very nice transition: How would you define uncertainty? And how does the model help understand uncertainty? Is there uncertainty in the model?

So, uncertainty: how do I define it? Well, you have to have a particular target for your model. If it's a weather model, there's uncertainty associated with the way you have to set up your model, the structural uncertainty, the parameters in that model, and the initial conditions in that model, all of those will lead to a different trajectory from where you think you're starting to the weather that will be present some time later. So, with these top-of-the-range models, uncertainty is associated with the multi-dimensional sphere around that point, which actually says this is…we can't say where within that.

And you have a sort of propagation and accumulation of uncertainty through the different phases of the model?

So, you have a trajectory, if you like, with this time-stepping, and there's gradually wider uncertainty around that. Now, for the simpler models, I don't think uncertainty comes in the same way, in that, given an idealisation of the system, there's uncertainty associated with whether you've done that process correctly, but I don't think there's uncertainty in the same sense. You have made approximations and those produce uncertainty in the idealisation you have made, and it's rather different from that sort of...

And in order to understand the uncertainty better, would you say that we can refer to what was said earlier about the sensitivity analysis? Is it just an aspect of it?

Yes, so, a sensitivity analysis almost becomes a crucial thing in the uncertainty at this end, in that you can look at the sensitivity, if you know the sort of things in the initial conditions and in the parameters the output might be sensitive to, then you can produce different runs of the model that explore those sensitive parameters or initial conditions, you can use an ensemble of those different runs which then you hope encapsulates the uncertainty in that sphere. But as we've talked before, you don't know the interaction between those, and still there is the problem of the structural uncertainty if you do this with one model, but then what is done is actually to look at how various different models from different centres do the same thing, and hope that those will conquer to some extent the structural uncertainty; and I know Lenny Smith, for instance, takes something I've said in the past which is if you take particular phenomena that occur in the weather and climate systems, then we know that all the models don't treat them very well; so, for example, there's a system that affects us: blocking, that's when a great high-pressure system sits there and it stops the weather systems coming in, and almost all models have under-represented the frequency of that sort of event. So, however much you do your ensemble or your post-processing or whatever, you can't get over the fact that you're not representing the thing; so if you did a post-processing, you couldn't say that in a different climate in the future, you can't post-process to say well how that thing that we don't represent will change in the future, we don't know. And so there are certain things that cannot be handled by that. So, there is always more uncertainty than you can represent from your ensemble, but we don't have a handle on that, it's the unknown sort of thing.

How do you define risk?

Well, I know these days they give a formula for risk of 'hazard times impact' or whatever; for me, risk, I do tend to think more 'what is the risk of…?' To me, it's in probability terms, what is the probability of a colder summer than normal, or whatever. For those who want to get more towards dealing with the impact as well, in some sense, you have that at the back of your mind, in that what's going to be an extremely hot or extremely cold summer, and what is the probability — how is the probability distribution changing for the summer? That very much to me sums up risk.

So, risk for you is something that can be presented in probabilistic terms, whereas uncertainty is not? You're an economist!

Am I? Oh dear!

That's really interesting, because this distinction, the one you're making, is the one that is made in economics, whereas it's not common…the variability on these two questions is massive.

I think, probably, it's because it's a very practical part of our subject, and so it's very much interacting with those who want to know things, which always inspires the research. I mean, the Indian summer monsoon — is it going to be stronger than normal or weaker than normal? Well, that's just an amazingly important question, and if you can answer it in any way, and there's no way it's a deterministic thing, all we can ever hope to do is say well, this is more likely, we know that given the chaotic nature of the system, the sensitivity to all sorts of things, we know it's only a biasing of the dice that we can look for.

And so how does the model help us understand the risk? Suddenly being able to produce a probability, this is quite amazing, so how does the model help here?

Well, this end of the range of models, this complex end, this ensemble that we've designed, we hope, will then produce an idea in the model world, in the world of our model, how has that probability changed for this summer? Because we have run our model many times and then we are running it for the particular situation that we have this year; and we can then look at this 100th member, or whatever, we can say look at how that seems to influence... Now, of course, you have to give a probability to each of those members, and usually what is done is they all are equally likely, and then we compare that with the probability of our... Now, of course, then you have to say, that *is* in the model world, you might then have some flavour that our model is typically biased in this way, but still...

It's probability with a disclaimer?

Yes, that's right, absolutely. And the trouble is, you see, what happens, which is what happened recently, I think the Met Office said there was..., ok there are five categories of temperature for the summer — much above, above, normal, below normal, and much below normal — and what they said was that there was a slightly increased probability of much above normal and slightly reduced probability of much below normal, and the headline in *The Mail* or *Express* was "It's going to be a sizzler!"

So that's why language is important!

That's right. So, somewhere along the line, and maybe wilfully, the actual nuances of what's been given have been lost. But *c'est la vie*, that's what happens.

The last question which I think is a more personal question: what is the experience or result in your field that has had the most significant impact on you? It can be something of your own research, but it can also be more general than this.

I saw this and I thought, I don't see there's been the revolutionary moment, not since I've been an active researcher anyway. I wondered quite what to put in this. Most of the things that are around have, you know, there's a basic paper that people would refer to but I think usually the knowledge has been incremental, and my own best work has been putting together things that existed but hadn't been put together before. Actually, I'm going to borrow one from my own work I think: there's one bit of work I did that people tended to look at something happening here affecting places to the east of that moving as the westerly winds come; now what I showed was that actually you can have things happening in the tropics and you get communication along almost a great circle path, so it's actually very easy... so I showed the mechanism by which something happening in the tropics can influence somewhere in middle latitudes by sending out a stationary wave that propagates from there, but *not* along a line of latitude; the winds may be coming from the west, but the communication can be like dropping a stone in a pond, you get ripples going out in all directions.

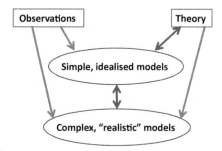

Chapter 8
A Conversation with Charles Manski

Can you describe your field briefly?

I do econometrics with primary applications to public policy.

Present a paradigmatic example of a model in your field, describing it in terms that are accessible to non-experts.

A very large part of the work is modelling decision-making. So, economists for a long, long time have had a basic model of maximising something — maximising utility for consumers, maximising profits for firms. Usually, we're modelling behaviour under uncertainty. The people making the decisions themselves face uncertainty, so they have to have models or assumptions about people's preferences, how they form beliefs, expectations about future events, and what kind of decision criteria they use. Economists put that all together in a mathematical model of decision-making, i.e. a model expressed in formal mathematical language. (Social scientists sometime pose models in verbal terms, without use of mathematics. This is common today in much of sociology, psychology, and anthropology. It was the norm in economics until the late 1930s, and it still occurs to some extent in economics.) Then economists try to estimate parameters of the model, of features from the data, and then use the model to forecast behaviour in new situations.

So, that's a classic problem; many, many economists are working on problems of this type and I too have, at various points throughout my career. As just one example, I was giving this masterclass yesterday and the day before here at UCL. A very important problem has been to study the effects of income tax policy on the amount that people work, on labour supply. So, we try to model how people allocate time to working and non-working. Hours are affected by salaries and by income taxes, and we try to use the model to forecast how behaviour would change. This is the really important point from my perspective — the use of the model, not modelling *per se*, but the use of the model to forecast how behaviour will change if the policy were to change. And that becomes an input which hopefully will help make a better policy.

In a nutshell, because models are very different in different fields: So, would you say that in the model, you have a set of assumptions, then some criteria — some decision criteria, and then from this — and data, we'll come back to data — you will have outputs of the model that you use for various purposes?

If you want an example, a particular class of models are called random utility models. They assume that persons maximise utility, but different people will have different preferences, and so you would mathematically model people across a population as having a probability distribution of preferences. Then utility looks to the researcher as a random variable. Random utility models began in psychology, not in economics, a long time ago in the 1920s. And there the model was not of a rational decision-maker, but rather of a person who doesn't know what they want, so every time they make a decision, they pick a utility function at random from inside their head, and that's where the random utility model came into psychology. But when economists began doing this formally, which is more recent — the late 1960s or the early 1970s — then the randomness was not inside anyone's head but across people. So, the three of us might have different preferences. The reason why that's important is that sometimes the same mathematical model shows up with different substantive interpretations across fields. I could actually give examples from some of the physics models which have flowed into economics as well.

Could you describe the different steps involved in a modelling process in your field (taking a particular example if this helps)? You already mentioned lots of mathematics or mathematical tools. What would you say is the role of mathematics in modelling in your field?

To prevent you from fooling yourself! I sometimes write in words, but I usually trust explicit mathematical statements much more.

So, would you say it goes beyond a tool? Or it helps to formalise…?

Oh, absolutely. It's very interesting — if you were to talk to an economist and a sociologist about theory and say what's an economic theory and what's a sociological theory, modern economic theory at least since the 1930s has essentially all been — hopefully — coherent, concise mathematical models; but if you talked to a sociologist *today*, some sociologists would use mathematical models, but some of them would be writing entirely verbal descriptions presenting broad theories of society. English or French or any language — all verbal languages — are kind of limited because any word gets defined in terms of other words. We all speak — I'm not speaking mathematics, obviously — we speak verbally, but it only goes so far if you want to really get underneath something and particularly, of course, to quantify something and not just talk in qualitative terms. So, we have to use mathematics.

For instance, coming back to the assumptions, would you say you think of them first more mathematically or more conceptually?

I think it goes back and forth. Of course, different researchers may work in different ways. For myself, you're thinking in your head and you have some ideas but then… I'm not exactly sure what's in my head and so I have to write it down in, you know, formal notation and hopefully you have a good concise notation… Because you can have obscure mathematical notation too and it makes things harder rather than easier. But sometimes in my own work, half the work is figuring out just what notation to use. And then proving theorems…

I fully agree, and I think also working with other people might also be difficult for this question of notation because even people from the same field, sometimes we have different types of notations and then it makes difficulties...

A long time ago, when I was Assistant Professor — so if you've worked on probability theory, you'll know what I mean — I had to set a formal notation for a probability space, and I wrote something like '(x, Ω, P),' so Ω was the sigma algebra, and somebody said to me, 'You can't use omega, you have to use capital sigma' — or vice versa, I don't really remember what it was — but he couldn't understand what I was talking about if I used the wrong symbol. Obviously, we understood each other fine in the end.

So, talking still about this mathematical formalism and notation, what constituents besides this mathematical formalism do you think are part of the model? For instance, the relationship with other aspects, like data, which you mentioned, at which stage do you see the importance of data? Do you have a first stage which is just your theoretical model and then you apply it to the data?

I think that's iterative too. I often hear in seminars, someone will say 'I have a theory and then I bring the theory to data,' as if you sit in a totally abstract way with the theory and then get the data later, and then use the data to test the theory. I think in practice it's much more complicated than that. If you're analysing decision-making, then we have loose anecdotal evidence on behaviour all the time. We do from even watching our own behaviour, or from observing basic statistics published from surveys and so on. And I think it really goes back and forth between the data and the theory. I think no one's really succeeded in formalising how people do move back and forth between data and theory. The closest attempts may be in Bayesian models, where you start with *a priori*, which is kind of theoretical, and then you add data and get *a posteriori*, and then you move on... But I don't think that the process is so rigid. As much as science has progressed, I think we're limited in understanding the cognitive process through which we combine data and theory. But somehow it seems to go back and forth.

I think it's quite important in some types of statistical analyses, you have this whole notion of hypothesis testing. R. A. Fisher was around here at UCL. Hypothesis testing has been very influential in some areas, but it's an extremely limited view of scientific work because what you do is you're supposed to form a hypothesis, which is the theorising, then you take it to data and then you test the hypothesis…then what? There's no 'then what?' because after you've tested the hypothesis, and you either reject or not reject the hypothesis, then what are you supposed to do? And even though this is taught in undergraduate statistics courses all the time, it's never explained. And then no one who does hypothesis testing just reports the hypothesis test and then goes home… Then they go back and iterate again, and then — to use technical terms — there're issues of pre-tests, or sequential testing… So, to summarise, the notion of just setting up a null and alternative hypothesis and then going to get data, and then writing up the results, this simple two-step sequence just doesn't exist, it's much too simplistic.

What would you say is the role of language, or the qualitative aspect, in modelling?

I spend a lot of time just thinking what word to use to describe something. I know one of the questions you have is the difference between a model and a theory…you posed it 'between a model and a theory,' those are two words, but you could have other words, a hypothesis, a conjecture, an assumption. There's overlap…but how people interpret these words depends on who they are. It's really a cognitive task, a psychological question, of how people use language, and then equally so how they communicate with one another. I have a book, and there was a very intentional thing that I did: The book is written 99% in English as opposed to mathematics, called *Public Policy in an Uncertain World*, 2013. I was intentionally writing for a broader audience, taking things that I'd done over the past 25 years and trying to write them in a more accessible way that a good journalist or a Masters' student in public policy might be able to understand, without mathematics. And I begin the book, in the first or second page, with one of the few, it's not mathematics but a logical relationship, and I put it in bold and I wrote:

$$\textit{assumptions} + \textit{data} \Rightarrow \textit{conclusions}.$$

It's obvious to anyone who does research, but it's very important… And I thought a long time about what words to use. 'Conclusions' was no problem, 'data' was no problem — I could have used the word 'evidence,' these are the synonyms, the verbal description — I thought data and evidence, they mean the same thing. But the question was 'assumptions': people said to me, well, don't you mean '*theory* + data ⇒ conclusions'? Why are you saying 'assumptions + data,' I said that I thought 'theory' was too grandiose a term, and that I wanted to avoid the word 'theory' because it seems too big, and 'assumptions' is a really mundane, ordinary word, and I didn't want to overdignify what people do by extensive use of the word 'theory.' I do use the word 'theory' in some places in the book, but basically for just broad frameworks, not for the ordinary kind of model. I could have just said 'theory + data,' or 'models + data' or 'hypotheses + data,' any of those. I chose to use 'assumptions' because it's such an ordinary word. So, I spent a lot of time thinking about that. Some people don't like that because they think it trivialises what they do, they would rather say 'theory + data' rather than assumptions.

Maybe the theory of the model is in the '+'

Well, that's interesting. I did use a + sign as a mathematical symbol…

There is a lot behind a '+,' there's a lot of science!

You're right, I hadn't thought about that. Yes, if you've got a relationship like that… What I meant by the '+' was the combination of the two; in words, you would say 'assumptions combined with data,' so I used the '+' sign, and I used the logical implication sign which I thought most people understand.

So, I think the words matter. One has to be very careful — there's a lot of spin.

Actually, I wrote a piece in, do you know, *Vox EU*? It's an online economics portal…it's more than a blog, they're a bit longer than Op Ed pieces; they're like little essays, 1500 words, on economics and public policy issues, and they do a very nice job, and have good readership;

I've written four or five pieces, and one of them was about the verbal semantics of discussions of income taxes and labour supply.

There are words that economists use to describe the incentive effects of taxes on work effort, which I came to the conclusion are very pejorative, they're very basically anti-government words — they talk about the 'dead weight burden' of income taxation, the 'inefficiency of the tax system,' and there's a whole set of words of that sort, which are entirely unnecessary — if you look at the mathematical modelling that's involved, but somehow these words have come into the literature. And I became quite concerned about the use of those words…economists when they use those words, they know what they mean, formally, and they just translate them into the mathematical modelling and don't really pay attention to them much; but when economists are writing for a broader audience, for the public, and the public sees the words 'dead weight burden' and 'inefficiency of taxes,' then it feeds the usual conservative view that taxation is bad. And that's inappropriate, that doesn't translate what the mathematical modelling means. So, I wrote a little essay that we should just get rid of using those words and just describe them in neutral terms. And this piece drew far more attention than any of the other four or five essays that I'd written for *Vox*; I had one person in the United States who has his own blog screaming that I was an idiot and that of course these words are essential. This guy is known as a conservative economist, and he really reacted very badly to this, but I got many more reads from raising this issue. So, I don't know whether in the physical and biological sciences the semantics matter…well, they may matter; I dislike how physicists make up incredibly grandiose terms, like the 'theory of everything,' cosmologists create a secular religion with names for it… Social scientists are prone to do this, but it's not just social scientists who are prone to choose particular words to make what they're doing seem more important.

That's very interesting because it's related to the use of models. Models are often said to represent a target system or part of the world. Does this apply to your field?

All the time. If you think about modelling behaviour, ordinary economic theory supposes that a person contemplates *all* of the incredible range of decisions that they face in their entire life, from what to eat for breakfast in the morning to who to marry, and whether to say this word or that word in a discussion, or which concert to go to. You're supposed to be simultaneously maximising global utility across all of these decisions. But of course, we don't know to what extent people in reality make decisions that try to cut across all those things — that would just be an incredible computation problem if nothing else. But certainly, in terms of the modelling of decision-making, economists pick one area, in a model of, say, labour supply. A labour supply model does not jointly model what you ate for dessert and so on. So, these are all what we would call 'partial equilibrium'…you have to put a boundary on the system. The term economists would use in English would often be 'partial equilibrium' rather than 'general equilibrium.' So, you take a piece [of the system] and you focus on one aspect and you essentially hold everything else fixed. So, 'hold everything else fixed' is a simple term; for a long time, economists would use the Latin term *ceteris paribus*, which just means 'holding things fixed.' So, you say when we're writing carefully, this is what we're modelling, *ceteris paribus*. You have to define some limits on the system because it's just impossible otherwise. Part of the science, or the art, of modelling is trying to figure out how to draw the boundaries of the system. If you can get the system to be smaller, then it's easier to do the modelling, but you might miss something. So, there's always a temptation to make the system larger, but then the modelling becomes more complicated and doesn't work out, so everyone has to cut it. Of course, this is true in physical and biological sciences as well.

You addressed this partially before, but what for you is the relationship between model and theory?

These are words that in my own mind overlap…if you were to analyse my own behaviour, if someone were to do a text analysis of things that I've written or have said in lectures, you could probably figure out what we could call my revealed preference from the choices that I make, you would probably see that I use the words 'model' and 'theory' in somewhat different ways. There's clearly an overlap. A model would be somewhat more precise and delimited; theory for me would be a somewhat broader set of constructs. So, I used the term Random Utility Model before: no one would call it 'Random Utility Theory,' it's never been called that. On the other hand, Random Utility Models fit within what is called Decision Theory.

And in Decision Theory, you might have different models?

Yes. Decision Theory is bigger. But then even within Decision Theory, some people say Decision Theory and some people say Decision Analysis, which is another term. So, we use these terms… Let me give you an example, not of those particular words, but to explain. I like to quantify things, but of course, like everybody else, I have to use words.

There's a whole branch of the work I do on measuring probabilistic beliefs. So, if we ask a very simple question: if someone is working and you want to know what they think is the chance that they'll be working next year, or that they may lose their job, or something like that. There's a long history in survey research done by sociologists and social psychologists, to ask people about their beliefs about future events, but to do it in words… And a very simple way is to do that as 'yes/no: do you expect to have a job next year?' But then they say no, that's too crude because there may be some uncertainty, so then they would say in words 'How likely is it that you will have a job — is it very likely, likely, not likely'? There's a major social survey in the United States that asks questions using that language. Back in the early 1990s, when I was starting to think about doing empirical research on this, I said I don't want to do that, because I don't know how people interpret the words. When one person says something is very likely, does that mean the same thing as if another person says very likely? Or even for the same person across different events — if I say it's very likely that the sun will rise tomorrow, does that mean more or less the same thing as something else is very likely? So I said, well, economists typically model beliefs as subjective probability distributions, so I decided to ask people what's the percent chance, on a 0–100 scale, that you'll be working next year? And since then, for not just me — lots of people do this now — that's the specific way that we do it, as a quantitative percent chance, because we were uncomfortable with the multiple interpretations of the words and the crudeness… So, that affects the way that you do your research. In terms of the difference between the words 'model,' 'theory,' 'analysis,' it's more vague.

What is the aim, or the use, of the model in your field?

If you ask different people who do the sort of work that I do, I think you would get two quite different answers to that. I'm clearly on one side rather than on the other. One group of people, which is *not* where I fit, basically say they're doing science, and the particular term that they use, that I really hate, is that they're trying to infer causality. The term 'causality' drives me crazy — I try not to use the word at all in my writing, except when I'm referring to someone else's work.

There's this enormous literature on analysis of treatment response — treatment A, treatment B, and what's the effect on some outcome. The language came from medicine of course, medical treatment: if someone's ill and you can treat them with surgery or with drugs, what's the outcome on length of life, and people would be interested in the average treatment effect across the population. And if you ask people why are you doing that, they say they are trying to learn a causal effect, what deeply is going on? And if you say why do you want to learn a causal effect, they say because that's what science is about. They say that science is about us learning how the world works, and distinguishing causality from statistical association is a basic question in science, so they want to learn about causality.

As I said at the beginning, I view my work as hopefully being useful for the formation of policy, I care about public policy, but it could be for firm decision-making or individuals… I want to provide information that might help someone. I don't care about causality for that, I care about predicting what outcomes take place if I were to make this decision or that decision. So, for me, the purpose of what I do is to predict counterfactual outcomes.

Does it include, in financial applications, what we would call sensitivity analysis? You modify *slightly* an input or parameter?

Well, if you want to go on to that, we could talk for the next two days about that! But for the past 25 years, I have done what has come to be called partial identification analysis — this is an attempt to do a very systematic form of sensitivity analysis, more than the little tweaking kind of sensitivity analysis. So, I think we really do need to talk about that, it has been central to my work for a long time. We'll talk about that more when we come to uncertainty and risk.

Are you using computer simulations?

Sometimes…as a last resort. … That's not to say it's bad — sometimes you have to do it.

So, what kind of relationship will there be between the computer simulation and the model — does it help refine the model? And what stage of the modelling?

Well, this too, in reality, is another iterative thing. It's related to what we talked about — defining the boundaries of the system — how complex you want it to get. Speaking for myself, and I think for most economists, we like to use models from which we can get analytical mathematical representations or predictions. Take, for instance, something that cuts across so many fields to set up an optimisation problem; if an optimisation problem is simple enough, then you can find a first-order solution — what solves the first-order conditions of this or that, and you get to see what's going on. But of course, if you set up a more complex optimisation problem, you can do it on the computer and solve it numerically. So, I think what I do, and what many economists do, is to first see whether you can get a model that captures things that you think are most important and still can get an analytical solution, and if that feels comfortable enough, you do that; but then at some point, you want to contemplate a more complex situation — it can be an optimisation problem or another important example would be modelling market equilibrium or equilibrium in game theory — can you solve it analytically or do you have to go to the computer, either for a deterministic solution or for simulations. So, for people like myself who do that, I would say it's an admission of defeat, but you really would like to get an explicit form, and you've got to go to the computer when you really decide you can't.

Now, seen over time, that's a big difference across fields. Computer scientists of course do algorithms, and they hardly ever think of getting an analytical solution — their first impulse is to go to the computer, that's what they do. And I think many physicists operate that way. A friend told me this story: there's a place in the States called the Santa Fe Institute, and back in the 1980s and 1990s, they tried to bring physicists and economists together to try to establish a common language and have communication between the two groups. Even though they're dealing with very different substantive problems, there's supposedly a common mathematical language. And the story is that someone set up some problem for them to think about overnight and come back the next morning and see if the two groups made progress on the problem; and the economists went back and tried to crank out an analytical solution and the physicists went back to the computer. They both have their place.

I had to deal with this very recently in a particular report — it will be public within a few weeks — I do a lot of public service work for the National Research Council in the United States, and the National Academy of Sciences, and also the Institute of Medicine which is part of the same thing; and this is a very specific public policy problem: the Food and Drug Administration has responsibility for regulating tobacco, they can set regulations on nicotine content in cigarettes, on advertising, etc. And the FDA was trying to get some help on forecasting the effects of different policies. And it's not something we could do by experiment, which is what the medical people like to do, because you can't force people to smoke or not smoke, so they knew they were going to have to use models; and what the FDA did is they gave a contract to a group called Sandia National Laboratories — they're a group of physicists who used to do nuclear weapon design, and they ran out of work, and decided that they would do social science instead; so they use what's called agent-based models — they go straight to the computer with some model of humans interacting like elementary particles or something which comes from physics, and do this agent-based model. And I was asked, my committee was asked, to evaluate their work, which we did not like at all, because they have no social science behind it, it's just maths without social science. So, we came down pretty hard on this. It's not published yet, but it's essentially almost there, so I think I can say that.

So, we have a committee that's made up of many different people, there was another economist, there was a sociologist, and some medical health people; but I and a sociologist were supposed to write up the chapter on basic principles of modelling and the use of different kinds of models; and we did this to, beyond our sponsor, or client, the Food and Drug Administration, to try to help public health people think through the uses of different models. And this question about when you do analytical models, and when you go to the computer, is something we spent a lot of time on and I hope we wrote up clearly. Within two weeks, I can send you this chapter when it's published. And we have sections in this chapter on the advantages and disadvantages of analytically solvable models versus the computer models.

We had one originally called simple versus complex models, but in the end, we changed the terminology to low-dimensional versus high-dimensional models because one of the reviewers thought simple and complex was too mushy, and we got into a semantic discussion in the review process — a reviewer questioned what does complex mean versus complicated — really! And we decided people use these words differently, so we went with low-dimensional versus high-dimensional. And what we concluded in the end was that no one approach dominates the others — the simpler, analytically solvable model is as effective as complex models which you can only solve numerically by simulation, that they both have their roles, and you just have to be careful about what you're going to do. This is a consensus report — we had 15 people on the committee and they all had to agree to the language, but I and one other person had done the drafting of this chapter, and I felt fully comfortable with what we wrote. So, we're hoping that this will be helpful.

Let me just say something more about this agent-based model, going back to this semantics question: it took us three or four meetings to agree that agent-based models and micro-simulations were technically the same thing. In my own mind, from the beginning, I always thought that these were the same thing: they just use computer code as the expression of models at a micro level of individual behaviour. Economists tended to use the word micro-simulation whereas the computer scientists tended to use the words agent-based modelling, and I thought if we look at the maths, these are the same thing. But we had people from different disciplines — the issue that you're facing — on my committee — economists, sociologists, operations research people, and so on — who would first view these things as 'no, no, no, no…'; someone wrote, 'Is it true that an agent-based model and a micro-simulation are the same thing — but that's the same as saying that a horse and a human are the same thing because they are both made of DNA,' and that it's trivialising the difference between an ABM and a micro-simulation to say they're the same thing. So, these communication issues that you're trying to — hopefully successfully — bridge very definitely came up in this committee, and I've spent a lot of time outside of economics talking to other people, so I think I'm somewhat more sensitive to these issues than economists who have stayed only within their own field.

What has been the impact of the development of new technologies or tools in your field? (e.g. telescope in cosmology, etc.)

The development of modern computers has been highly important to economics as it has been to other sciences. I would distinguish two distinct uses of computers in economics. One use, fundamental to modern empirical research, is to digitally store datasets that characterise consumers, firms, and aspects of the operation of the economy. The other use is to numerically solve complex optimisation problems and systems of equations and inequalities that characterise economic activities.

Last question on the modelling side, before we look at uncertainty and risk: what is for you a good model?

The word 'good,' you know, …other than saying I know it when I see it! It's a very hard question! And it's inherently, even though we're talking about doing science and being mathematically objective, 'good' is a very subjective term. When we talk about uncertainty, I'll start using the word 'credibility' quite frequently, so for me certainly a necessary condition for a good model is that the assumptions that it embodies have some plausibility, some credibility. Don't ask me to define *that* word! But then of course it could be a credible model, but not have any usefulness, so it has to have some power — a good model has to have some ability to make counterfactual predictions, to tell you something about the world that isn't in the data that you currently have. So, I guess a good model is something that has the power to make…the term 'counterfactual predictions' is very important, in a new environment, it is extremely important, but then they also have to be credible, believable, predictions. So, I guess that's what a good model is. I would say that of an economic model and I would say that of Newton's Theory of Gravity — although they call it the *Theory* of gravity, not the *model* of gravity — we now know, since Einstein, that it's not a perfect representation of the world either, but it's still a pretty good model for dealing with most things that you experience, even including space flight.

Now, we move to uncertainty and risk. So, starting with uncertainty, how would you define uncertainty (you can define uncertainty versus risk if you prefer).

Yeah, I don't want to take one value… I have to look at the whole constellation. Again, this is one I teach regularly, it's become extremely important from what I call the research perspective, and of course, many many people spend lots of time on these things, and there are different semantics too. I think the easiest way is to think about the maths, then we can attach the words.

A starting point for almost anything mathematically is to specify what's typically in economics called the state space — it's a set of all possibilities. The specification of a state space is subjective. You say what are all the things that are in the system, what are all the things that I think might possibly happen; I don't have complete knowledge of the world, so we're dealing with incomplete knowledge, so I list all the things that might possibly happen. Of course, there may be some other things that I've said, 'no, that's impossible' — it's logically impossible, it's not going to happen, and that's outside of the state space. But I have to start somewhere because I think it's clear if you try to do any formal analysis of uncertainty with incomplete knowledge, if you just say 'anything can happen,' you just get nowhere. I don't know any treatment of risk and uncertainty or ambiguity that doesn't begin with a specification of a state space. So now, you've got this list of things that might possibly happen; then there's a big question about, among those things that could possibly happen, what further can you say? The big divide is between putting a probability distribution on a state space of things that can happen and not putting a probability distribution down.

Now, within the research putting a probability distribution down, there're two branches — there's one that says, 'I know the probability distribution, there is some objective process, stochastic process, that's generating outcomes and I know what it is.' That's what economists usually call risk. There's a real-world probabilistic process and I know what it is, so we talk about decision-making under risk. A synonym that came out of macroeconomics is to say that I have rational expectations. The word 'risk,' I don't know when it was used...well, when Von Neumann and Morgenstern in 1944 (*Theory of Games and Economic Behavior*, Princeton: Princeton University Press) talked of decision-making under risk and rational expectations in the 1960s, in either case, there's an objective stochastic process and I know it, that's the way I use it. (There is some semantic confusion, Abraham Wald, when he talked of decision theory in 1950 (*Statistical Decision Functions*, New York: Wiley), used 'risk' to mean something entirely different, so I don't want to mix that up, that's not the way people typically use 'risk' today.)

So, now let's say you're willing to put a probability distribution down, but you don't think it's necessarily objectively correct, so then we have this whole subjective probability distribution that's called Bayesian analysis. So, then we put a probability distribution down on all the possibilities, but it's one that's in your head, it's clearly a subjective concept, and it's my personal probability, there's no objective process, I don't have some stochastic model for the outcome of the UK elections in six weeks or so, but I might have a personal probability that the Tories will win or that Labour Party will win, one's probability is 0.6, the other's probability is 0.4, and that's the probability in my head, so it's a subjective probability. Economists very often assume the first kind of risk they actually know, the objective probability distribution; when they don't feel comfortable with making that, that's a very strong assumption, then they'll say, well, people act as if they have subjective probability distributions, and we have this typical way that economists model behaviour so that people are maximising subjective expected utility, which is that they put a subjective probability distribution down on whatever they don't know and they maximise the expectation...

But then there's the question of what do you do if you don't feel comfortable putting any probability distribution down. Say there are five possible things that may happen and they're all possible, but I don't feel comfortable subjectively putting any probabilities on each one happening. In the areas where I work, the typical word for that has come to be *ambiguity*. Just knowing that there are these five possibilities, but not having a probability distribution. There are different terms for that across fields; the term 'ambiguity' is associated with a particular article by Daniel Ellsberg in 1961 ("Risk, Ambiguity, and the Savage Axioms." *Quarterly Journal of Economics*, 75, 643–669), which has become a very famous paper. Much earlier in 1920, there were two pieces of work on this problem: one by an American economist named Frank Knight, and macroeconomists following him talk about *Knightian uncertainty* — so there's not just one uncertainty, there's a particular kind of uncertainty they call Knightian uncertainty. I know from talking to macroeconomists that that's the term they use. It turns out that John Maynard Keynes also wrote on that, he wrote a treatise on probability, it never got called Keynesian probability, but he wrote on that same conceptual distinction between knowing and not knowing probability distributions, also in 1920. But people weren't really following up on that work for a long time until very recently. I've learnt that people in other fields use different words. I was at a conference a couple of years ago on "Regulation under deep uncertainty," and it turned out that **deep uncertainty** was the same thing — no probability distributions. I've talked to a geologist and some other economists who talk about model uncertainty. So, semantics really vary quite a lot across fields. But the maths really helps here because when you say 'well, what does that mean?' and it turns out that in every case it means not having any probability distributions.

My background makes me closer to your particular distinction. We recently interviewed a civil engineer and for him it was just the opposite — what we call risk would be uncertainty for him. So, it's quite interesting.

Yeah, so that's why the maths helps define what we mean by these things. So, those are the basic distinctions that I would make. My own work has focused, for the last 15 years, very heavily on ambiguity.

And in the modelling phase, or in your models, where would you handle risk or uncertainty? Depending on where you are — is it in your criteria, or is it at the assumption phase, or in the treatment of the output…?

Maybe the easiest way to describe it is to explain a little bit more about the kinds of things I do. Another aspect of this question, particularly in terms of empirical inference, is a distinction which is important, not just in economics but in statistical theory as well and critical in econometrics, when considering inference on a population, to distinguish between the problems of statistical inference and what we call identification problems.

Statistical imprecision and uncertainty everybody understands — there's a population, you want to learn about the population, but you don't observe the whole population, you observe a finite sample of the population, it may be a random sample to make it easy, and then there's the question of what conclusions can I draw about the population from the finite sample evidence? So, that's what statistics is all about. And we have all kinds of statistical theory about what happens about characterising statistical imprecision, confidence intervals, standard errors, and so on; it asks what happens if the sample size goes to infinity. And if the sample size goes to infinity, the statistical problem goes away — the sample average converges to the population mean, and everything…it's nice. But as the sample size goes to infinity, you get a better and better picture of the population. What identification is about is problems that don't go away with sample size.

For identification, you said you'd call yourself a probabilist on the Continent, so when we teach, we make this distinction between probability theory and statistics. Probability theory is probability distributions on populations and statistics is about using samples. Identification is applied probability theory: it's about, given the nature of the data that I can collect in the sampling process, what could I learn about a population, even if I could collect as much data as I want — if I could let sample size go to infinity — there may be some things I can't learn. Those are called identification problems. … When there's an identification problem, when the kind of data that you have is really sort of weak, and it's just not going to let you learn a lot about the population, the traditional answer by economists to this would be to combine the data with strong enough assumptions about the probability distribution of interest so that you can get exact conclusions, at least as sample size goes to infinity, so you get exact conclusions. So what happens in econometric practice is that we define what we want to estimate, and we'll get point estimates of the quantities, and they'll be consistent — they'll converge to the truth as sample size goes up.

What I found — and in retrospect I realise that this has always been on my mind for my entire career, but came to the front in the late 1980s — was that I often felt very uncomfortable with the assumptions that people were using in order to draw those conclusions. So, I decided to back off and say, 'well, what conclusions could I draw with weaker assumptions?' And if you use weaker assumptions, economists would typically say, 'well, it's unidentified and you can't learn anything.' But I realised that mathematically that often wasn't true — that you might be able to learn something, just not the whole thing, it's just one more abstraction and I'll give an example: you may be able to learn that the truth lies in some set rather than the truth has a particular point value. That's what has now become known as **partial identification** — doing empirical inference that works with combinations of data and assumptions such that what you can learn about a population is something about it, but not the whole thing — you can get a bound or an interval or something. I've written so much on this for different audiences, but one I wrote particularly aimed at mathematical statisticians was a 2003 book in the Springer series called *Partial Identification of Probability Distributions*. It's a more mathematical treatment than some other things, but it says that what you can infer using statistical theory and doing the asymptotics is that you'll learn that the true probability distribution lies in some set of distributions, in an informative set, in which some distributions are possible but not others, so that's partial identification of probability distributions. That's a very formalised notion of sensitivity analysis.

Could it also be seen as a way to reduce ambiguity or to control...?

Yes, it's a way of characterising, that's exactly what took me, when thinking about decision-making, to ambiguity, because what you can learn from the empirical analysis is something that lies in the set, but I couldn't put a probability distribution inside the set, you just know it lies in the set, so it naturally came into ambiguity.

Let me give a very specific example, it's the one I began with in the late 1980s, and I think it's still the easiest one to analyse: say I'm doing a survey of the population and I want to learn about some characteristic, let's say median income of households in the United Kingdom. Anyone who does survey research knows that there's missing data — some people just don't answer the survey at all, it's a very practical question, some people will answer the survey, but will refuse to answer certain questions. Among Americans, income questions are particularly sensitive, we get high non-response rates, we get 20%, 30%, or more % of people who refuse to tell us about income. So, what are we going to do? The government will put out statistics on median household income, whether it's here or in the United States, how do they do that with 30% missing income data? The tradition has been to make quite a strong assumption, which is that the missing data are missing at random. The assumption that the data are missing at random, mathematically, is an assumption that the probability distribution of the missing data is the same as the probability distribution of the data that you have. So, then you can use the data that you have to produce a point estimate of median household income, or something like that. So, you make the strong assumption, and even though there's this basic inference problem that a lot of the data is just not there, you use imputations, you can do sample weightings of response distributions, but they're all based on the assumption that the data missing are random.

For a long time, many economists have been very sceptical of the assumption that the data missing are random because economists tend to see this as a decision problem: should I answer the survey question or not? People make calculated decisions — do I want to tell this researcher my income or not? And that's when you think it's not random — that someone who's wealthier may not want to tell their income more than someone who's just got an ordinary job and says who cares? Someone who has criminal income... So, there can be very non-random patterns. Since the early 1970s, economists felt that they had to have an exact estimate, but they didn't like the assumption that the data missing are random, so they came up with models of how people respond to surveys. These models are alternatives to assuming the data missing are random, but they also generated exact predictions, or at least probabilistic predictions, and so they can get point estimates. Well, I think those models are not very good either.

So, when I started working on this in the late 1980s, I wasn't happy with either approach. Missingness at random is a statistical assumption, so if you should be interviewing, say Donald Rubin (a statistician), he would suggest to use that approach. He has pushed this his whole career and has been very influential; or if you were to interview an economist like Jim Heckman, he would say no, we model behaviour. My own view is to be much more conservative. So, where I started was to say, well, what if we knew nothing at all about the people with missing data? Let's say there's 30% non-response, and of the 70% who do give us the answer, let's assume the answer is correct, and not worry about inaccurate answers (which is another issue), so there's 70% for whom we do have an answer and 30% for whom we don't. Then for 70% we do know the truth and for 30% we know nothing, and so if I want to calculate, let's say, the fraction of people who are below the poverty line, the American poverty line for a family of size 4 might be like $25,000 a year or something, well this is a very simple idea: among the people whose data are missing it could be that all of them are poor or it could be that none of them are poor, or somewhere in between.

Well, if you just think about that, then it's immediate that what you're going to get is a bound, an interval, you'll get a lower bound on poverty if all the non-response is from high-income people, and you'll get an upper bound on poverty if all the non-response is from low-income people, and you know for sure the truth is in between the lower bound and the upper bound. So that's giving you a set. That's a set with no assumptions at all about the nature of the non-response. That's the very earliest thing that I did ('Anatomy of the Selection Problem,' *Journal of Human Resources*, 24(3), 1989, 343–360), suggest that people produce interval estimates like this. What's happened over the past 25 years of course is that this kind of work has gone on in many, many directions; you don't want to just say I know nothing at all, maybe I know something, maybe I know something about the way people choose to respond or not respond, how that's related to their actual incomes so that I can narrow the bound, it's actually like you control the ambiguity. What you can learn depends on what assumptions you make — if I make a strong enough assumption, I get a point estimate; if I make a weaker assumption, then I'll get a bound, but maybe it's a narrow bound, if I make it a weaker assumption, I'll get a wider bound. So what I've been doing for a long time now, and people that do this kind of work, is to entertain assumptions of various strengths and say well what can I learn, let's say the data are whatever they are, I don't have the opportunity to collect new data — what can I learn if I combine the data with this assumption, or with this assumption or that assumption. Now that sounds like sensitivity analysis; that's what happens in sensitivity analysis — you tweak the assumptions…

That's an interesting one because in many disciplines, sensitivity analysis is a tweak in the parameter, but what's here is more the assumptions themselves…

Yes that's why I don't use the term sensitivity analysis, because I agree with you about the usage of the words. With sensitivity analysis, someone starts with 'this is my central model' and then does some tweaking… So I think the way people use the term sensitivity analysis is this tweaking of a parameter of a central model; and I very much — because I was not starting with a central model, I was rather starting at the opposite end of having no model at all basically, with very weak assumptions and then trying to study the system…so there'll actually be a whole continuum of models.

This is really interesting because I think this is the first interview we have done so far where the assumptions have such a crucial role in the model. For the other interviewees, in some cases we didn't really spend much time on assumptions.

For me, it's first and always the beginning. There have been views in economics which are very different. One which was very influential for a while, I think less so now, is associated with Milton Friedman from the 1950s. Friedman was very famous as a macroeconomist, but he also engaged in philosophy of science, and he argued in an essay in 1953 (it became quite well known in some circles) that assumptions don't matter, all that matters is how well the model does in practice. He said you can make stupid assumptions, but if it works well, fine. What he never explained is how you would know that the model is working well if you can't observe the accuracy of predictions of counterfactuals... So I never ever liked Friedman's view; but I have now for a long time thought — and I'm not sure if this shows up in your interviews — that some people don't like to concentrate on assumptions, because when you do, you realise well, where are they coming from? And so Friedman says 'Oh don't worry — make whatever assumptions you want, as long as you can argue that the model is working, then the assumptions don't matter.' I just don't think it works that way. Anyway, the missing data problem is a very easy one, and in many cases like that now we have a distinction between the way I would use sensitivity analysis...and I think the way you do sensitivity analysis... If someone says the kind of thing I do is sensitivity analysis, I say, 'well, I don't mind that you use that terminology if what you have in mind is a very broad systematic sensitivity analysis; if what you mean is just looking at a parameter or tweaking, then that's not what I'm doing.'

It seems closer to what people in engineering or finance call stress testing. Under various assumptions, pushing to more extremes,...

Until it breaks down.

...like scenarios, of things like this.

Stress testing is interesting. Of course that's an engineering term originally, it's come into finance since the recession I guess... I think that's a good idea. There are occasions, it hasn't been that prominent in my work, but there are occasions when I've done something similar and I've used the term identification breakdown — you start out with the strong assumptions, and then you weaken, weaken, weaken and you move further and further away until you can't learn anything at all. There are various literatures that are related.

In statistics, there's a literature that began in the 1960s called robust statistics, and I could explain the similarities and the distinctions with the identification work that I do, some are likes and some are differences.

In the robustness work, the notion of a breakdown point is very important, so stress testing — how far can you push something before it collapses — I think that's good; and in fact, when talking with students about sensitivity and someone mentions sensitivity analysis — in fact, this just came up yesterday in the masterclass, someone used the term sensitivity analysis, I said I'm often very sceptical when people say they're doing sensitivity analysis, because when people report sensitivity analysis, their conclusion always says 'I did sensitivity analysis and it didn't matter — my result still works' or 'I did sensitivity analysis and I'm still ok'; and I told them if someone does that, that's a dangerous signal, because if you're doing a serious sensitivity analysis, you should push the sensitivity analysis until the result which you think is good breaks down, which is exactly the stress test, because you need to find how much you can modify the parameters or the assumptions for the result to breakdown; if you haven't found that, then you haven't done a sufficiently serious sensitivity analysis. So I think this stress test idea, that's a nice operation; and again I might use the words stress test.

But where this definitely comes up mathematically — so we've talked about missing data problems, and we've talked about analysis of treatment response where we want to learn an average treatment effect — is treatment A better on average than treatment B — and you're looking at how well it works on average in a population — so then a critical question is, is the average treatment effect positive or negative? Because if it's positive, I know one treatment is better than the other, if it's negative, I know the other treatment is better than the first one, if I don't know the sign of the average treatment effect, then I don't know whether treatment A is better or worse than treatment B. So then the clear thing for the stress test is what assumptions does it take to get the sign of the average treatment effect — if I get a bound or a set that covers both sides of zero, then your ability to discriminate the better treatment has fallen down, and you want to learn that.

So, now our last question, which is probably a more personal question: what is the experience or result in your field — not necessarily from yourself, but from research in general in your field — which has had the most significant impact, and why?

I'd have to talk about two different stages of my career: the second one is when I started doing the work that I've just described — because it totally changed, very broadly, the way I think about things. I, like every other economist, was starting with this missing data problem...and that you always have to work to get a point estimate, and when I realised that that's not necessarily the case — would something be more believable, be it a bound or a set of values, an estimate? Then I began to see the entire world that way.

You unlocked something?

I unlocked something. It's clearly the most important thing that I've ever done, and without being arrogant about it, and it's not as well known as it should be, but still it's had some influence. I've had a whole sequence of very good students who are now senior people at other universities and doing this kind of work, and it's spread among many other people; there's a lot of this work right here in this building here at UCL — Andrew Chesher and Adam Rosen (Adam Rosen is a Reader here now, he was a student of mine from Northwestern; Andrew Chesher is an old friend) and they've taken these kinds of ideas in their own directions. I think it's the way that everybody should be doing research. It hasn't got, by any means, to have that amount of influence, but it has permeated the economic research community enough that people think about well we don't have to make these strong unbelievable assumptions, we can explore what happens with weaker assumptions and see what we get. It led to all this work on ambiguity and decision-making, and has totally changed what I do.

So that was like in the late 1980s. It came out of a very practical question that a friend at the University of Wisconsin asked me: it was a guy in social work — this is about how the data and theory interrelate. He was a very applied researcher I knew more personally than professionally, who was studying homeless people in Minneapolis, Minnesota, and in studying homeless people, there's missing data — it's sometimes hard to find them — he interviewed them one month and tried to find them six months later or so; and he asked me, 'what can I do' — he said, 'there are these things in the literature about missing at random I don't believe that, there's these economic models about missing data, I don't believe them. What can I do?' And I thought well what can he do if he doesn't believe those things?

The earlier experience was at the very beginning of my career, when I was a first year PhD student at MIT that I was extremely fortunate because I had come into — this is the earlier thing that affected me strongly — that I had come in with a very applied interest in modelling choice behaviour, and I was particularly working on a model of college-going behaviour, and I later wrote a book on this (*College Choice in America*, with D. Wise, Cambridge: Harvard University Press, 1983). And I didn't know how to do it, because the standard economics for modelling behaviour was for modelling like the demand for apples or oranges — how much does a consumer demand, how much do you want to buy of this or that, and [choosing] college is like a zero or one decision: I either go to UCL or I go to LSE or I go to Cambridge, it's not how much of something it's which one do I want to do. There was no framework in economics, in retrospect it seems…but at that point, it's now over 40 years ago, there was no framework for doing that.

When I was a first year student taking my initial courses, there was a visitor, Dan McFadden from Berkeley, and he had just begun developing his random utility models that I mentioned at the beginning and developing what became known as multinomial logit, and conditional logit analysis, which is the same thing, and he became very famous for this — he got a Nobel prize for this work later on. I just, by happenstance, was there and I sat in on this course, and I thought oh that's the way we should do this. His work wasn't even published at the time; and for the first 15 or so years of my career I did work of that type, I still do some work like that, but it's not what I do mainly now, but it totally changed… And I think Dan, it was clear he changed the world view in empirical micro-economic analysis and choice behaviour, there are simply tens of thousands of applications of these discrete choice models. So that's the other thing that really affected me. I actually wrote about these questions, what's affected me, how I go about research, I'm not embarrassed by this — I put it on my webpage so I can't be embarrassed — there's a magazine or a journal for undergraduate economists called *The American Economist*, and occasionally they'll ask someone when they get to a certain age to write a kind of essay or something on what makes you an economist.

I think it can be very motivating for others.

So I did, and I decided that there were some lessons — everyone has their own, we each go our own way — but I decided…so in terms of what we've been talking about, I gave it a title I like very much, I called it *Unlearning and Discovery*. And unlearning is a necessary…you know when you're in school you're taught here's the state of the art and we should add to the state of the art. But anyone will tell you that I'm a very sceptical person, and so often what I find may be uncomfortable with the state of the art, and often what I need to do is to say, 'well, why is this the received theory or the conventional wisdom? Does it really hold up or not' — just to view it sceptically. And sometimes you have to decide no, it's not right, and you have to unlearn it, and sort of go back to basics. So I wrote both of these up as prime examples about unlearning the traditional way that economists have done consumer behaviour, and unlearning the way econometricians thought about estimation. I also wrote up, which we haven't talked about, whether you get your results easily and you publish a paper and it's successful — it's not like that at all. When I went out on the job marked after my PhD, almost didn't get a job because no one else was doing this work at that time, and on this work on ambiguity and partial identification, although I could get the stuff published in the 1990s…there was a *lot* of push back, an awful lot of push back; even to this day in some circles, although there is much less than there used to be. So part of what I was writing for younger economists was that, well you don't want to be stubborn, but on the other hand, if you think you really have something, you have to have enough perseverance to stick with it.

Chapter 9
A Conversation with Chris Impey

Can you describe your field briefly?

My research field is observational cosmology, which is the study of the size and shape of the universe, and my speciality is how galaxies and the supermassive black holes they contain have grown and evolved over cosmic time.

Present a paradigmatic example of a model in your field, describing it in terms that are accessible to non-experts.

The model I'd like to concentrate on is the biggest picture of all, which is the model of the universe — the expanding big bang model, because that underlies all of cosmology.

So, using that example, can you explain why a model is needed and what this model is?

Let me describe the model briefly without jargon. The model we have of the universe is an expanding space–time with an origin 13.8 billion years ago, within which we find all the matter and radiation in the universe. Our estimates of the contents of the universe are about 100 billion galaxies, and an amount of matter that is dominated by radiation — there are about a billion photons for every particle in the universe. Those are the contents.

The container is the space–time which has a distance in any direction from us of about 46 billion light years.

So, why is a model needed? Because going back through the long history of cosmology from ancient Greek times, people have speculated about the size and shape of the universe, but they had no tools to answer the questions, so it was pure philosophy. Even 25,000 years ago, one of the most interesting questions was: is the universe finite or infinite, because that's a very profound distinction. In a very intriguing way, the fundamental issues for cosmology were framed by philosophers two millennia before the telescope was invented, and the questions haven't really changed. Even at that time, philosophers recognised this dichotomy, that if the universe has a boundary, or an edge, what's the nature of that edge? And that begs the question what's beyond the edge? Then, if the universe is infinite, what does a boundless universe mean? To philosophers who were also dealing with the mathematics of infinity, an infinite physical space seemed to be a strange concept. So right away you've got an issue there — on either side of the dichotomy you have something that you've got to deal with. Astronomers are now using telescopes and sophisticated mathematics and modelling, but it's the same question.

May I interrupt you for a second here, because I think you feel — and you are quite different from the other experts so far — it seems that tools, like telescopes, have also played a huge role, the development of tools, in order to move the model a bit further. So big steps seem to be linked to specific tools — is this true?

That's right. The development of the tools was very important. Going back to the origins of the subject, which I would say would be philosophy 25,000 years ago, approaching the universe through pure thought has value and merit, and people have always been doing it. But astronomy and cosmology are empirical subjects, so it's assumed that you can only advance knowledge by making observations. This leads to another debate, another dichotomy if you like, as to whether or not you can understand the universe through pure thought or whether you have to make observations. This debate goes back to Plato and Aristotle, and I like the fact that my subject, which is so modern with the Hubble Space

Telescope and incredible tools, is still cast the same way as it was at the beginning.

The tools transform everything, because in the early history of naked eye astronomy there are no tools — our tools are the naked eye, and the development of simple mathematics, and understanding how light works. Even before the invention of the telescope, there's some sense that light moves through the universe, it reflects off objects like the Moon, and so you can understand the geometry of space. There was speculation, and a very simple geometric argument in the Greek times for a Sun-centred Solar System. It was outweighed by Aristotle and so the geo-centric model dominated for two millennia, but the logical argument for the Sun being larger than the Earth and the Earth going round the sun was made by Aristarchus, but there were no tools to make that model stick. So in ancient times it was a small universe model.

The telescope is the first transformational tool of astronomy. When it was invented in 1610, you had the ability to put the third dimension into space, and so Galileo and people who used the telescope very quickly cemented the Copernican idea and now you know that the Earth is not the centre of the universe. Immediately the universe is much larger, but you cannot put an upper bound on the size of the universe because it turns out that the distance to the stars is hundreds of thousands of times more than the distance to the nearest planet. While you can map out the Solar System, you're unable to use geometry, via the parallax method, to measure the distance to stars, so it's still just a matter of speculation. However, the universe is instantly hundreds of thousands of times bigger than in Greek times with the invention of the telescope.

Through the first two centuries of the telescope, tools improved, the telescopes got bigger and bigger, with William Herschel and others, and measuring distances to the stars and the mapping beyond our stellar system continues, but you also have an obstacle because your tools are just not good enough. It was not until the 20th century with Edwin Hubble — he benefited in his career from three times using the world's largest telescope for the first time — that you can demonstrate that some of these fuzzy patches of light are galaxies, distant stellar systems. Hubble's observations expanded the size of the universe by factors of thousands, again because of new, better tools. So yes, the model of the universe, whether it's a small Earth-centred universe or a larger universe that is basically a big galaxy, or a universe of galaxies, is all driven by the tools. There's a second layer of the tools involving spectroscopy. Unless you have the ability to disperse the light into a spectrum and therefore measure the motion of objects, you don't have the ability to understand that the universe is expanding. That's a foundational step, also taken by Hubble in the 1920s, to show that the universe is expanding, and measure for the first time its expansion rate and its size.

And what is the role of mathematics in the model? You mentioned briefly a mathematical model in the Greek times, a bit of geometry, but there's more about mathematical formalism or quantitative methods, which I think probably is quite relevant?

Mathematics is essential and foundational in our understanding of the universe in modern cosmology. Contemporaneous with Hubble's demonstration that the Milky Way is not the only stellar system and that the galaxies are moving away from us, there is the development of general relativity. This new theory of gravity from Einstein was quickly recognised as a brilliant innovation, and a brilliant theory, and also it soon had observational affirmation. It's very unusual for a major theory to get experimental confirmation within a couple of years. So at the same time, you're discovering that the universe is more than the Milky Way galaxy and that the galaxies seem to be receding from us, you have a physical theory for gravity which is profoundly different in concept from Newton's theory of gravity. Newton's theory had been working very well for hundreds of years, and people forget that Newton's theories still describe most parts of the universe extremely well. It only fails in situations of strong gravity or in the global situation where space–time could have curvature in general relativity.

This mathematical modelling through the physical theory of general relativity is critical to making progress in cosmology. It's interesting that Hubble himself, who was not a theorist, is largely unfamiliar with it. The theory of relativity had been published a decade before he made his critical observations, and Hubble never actually used the phrase expanding universe, he was scrupulous as an experimenter and he just described what he observed which was a recession velocity. He said the galaxies have a velocity moving away from us — a redshift as he measured it, and almost none of them had a blueshift. He chose not to interpret that, because he was not familiar with the physical theory that would have let him interpret the redshift in terms of expanding space–time. The interesting historical story that goes with this is that Einstein kicked himself; he famously wrote that he essentially missed predicting the expanding universe. This happened because when he visited the Mount Wilson Observatory, the place where Hubble worked, soon after his theory was published in the late 1910s, he knew he had a mathematical model for the entire universe, and he asked the astronomers 'what's the state of the universe?' and they basically described the big Milky Way model of the universe where we live in a vast stellar system that's probably hundreds of thousands of light-years across. The astronomers said that within that big galaxy the stars have mostly random motions with some rotation, but no net global motion. To Einstein, that meant it's a static universe — the galaxy has always had this state and it's not going to change in the future.

However, Einstein also knew when he solved the equations of general relativity for the first time that they're dynamic equations, their solutions involve either expansion or contraction. If you solve them for a force that happens in a space–time with matter in it, it has to be dynamic. So having talked to the astronomers who said the universe is not dynamic, the universe is essentially static, he suppressed what would be the natural result of his equations with a term which became called the 'cosmological constant.' He later described this as the biggest blunder or mistake of his career, because he listened to the astronomers who had not yet discovered the expansion instead of 'listening' to his equations. This is a famous story in cosmology. General relativity applied to the universe implies either expansion or contraction, and within a decade of the theory being first published by Einstein, the German mathematician Friedmann and a number of other physicists and mathematicians had worked with the theory to describe expanding space–time That foundational work from the 1920s is still the basis of modern cosmology.

So, if I understand, you would say that this role of mathematics would be the quantitative argument that's used in different fields, like physics, that would have an impact on even the concept, and open new ways of thinking, on cosmology?

Yes, very new ways. If you start with the foundations of gravity and move forward to how you calculate, Einstein's concept of gravity was completely different from Newton's concept of gravity. Newton had gravity acting at a distance; he didn't speculate as to the nature of this action and how it could operate instantaneously over the vacuum of space. He famously said 'I frame no hypothesis.' As brilliant as he was, he didn't know. He had a theory that accurately described orbits in the Solar System, and predicted orbits of galaxies around each other, and things that were never discovered in his lifetime, and the theory still works very well. But it has some conceptual issues that imply that it may not be an ultimate theory, and these problems were even known at his time.

Einstein was looking to resolve some issues, some flies in the ointment, with Newtonian gravity. The critical problem was the fact that the orbit of Mercury was not well described by Newton's theory. In physical theory, you're not allowed to have an exception, you've got to be rigorous, you've got to understand everything. Technically, the unexplained phenomenon was the advancing perihelion of Mercury in its orbit of the Sun. Mercury being the planet that experiences the strongest gravity, if there was any departure from Newton's theory, that's where you would see it the most. By the early 20th century, this was recognised to be a serious problem, that Newton's theory failed when trying to explain the orbit of the innermost planet.

Einstein went back to the drawing board in thinking about gravity, and what he came up with was a geometric theory of gravity. General relativity is rooted in geometry, which is another echo of how the Greeks had approached thinking of space and time. Einstein developed his own mathematics to make this geometric theory of gravity. He reached back, very interestingly, to mathematics that was developed in the mid-19th century by mostly Russian mathematicians — Bolyai, Lobachevsky, and a few others — brilliant mathematicians who were working on multidimensional spaces. They were trying to extend Euclidean geometry, not only to the situation where it's non-Euclidean but also to the situation where it has more than three dimensions. In terms of pure mathematics, space can have an arbitrary number of dimensions. So Einstein used multi-dimensional mathematics and the possibility of curvature, and he made a physical theory where space and time were coupled. Newton believed space and time were linear, they were infinite, they were uncoupled, and completely independent of each other. In Einstein's formalism, space and time are linked mathematically. General relativity relates the mass–energy of the universe to the curvature of space–time. These are the ingredients of his theory.

The core mathematics, which is designed to scare graduate students in astronomy now — they all have to take a general relativity course — is a set of 10 second-order, couple, partial differential equations. You can pack them down into one neat little tensor summary equation. Mass and energy are of course linked by $E = mc^2$, and space and time are linked through Einstein's theory. The tensor form is the most efficient packing of the equations, but when you want to actually do a calculation, you have to break out the 10 equations and do some challenging mathematics. It was recognised immediately that this was a brilliant formalism. Subrahmanyan Chandrasekhar, one of the greatest astrophysicists of the 20th century, called general relativity the most beautiful physical theory ever and he talked about it in purely aesthetic terms. He said that the mathematics behind it was beautiful, and the physical concept of gravity was beautiful, and very alien from Newtonian theory. Since Newton's theory still worked in the situation of weak gravity, Einstein's theory has to reduce to Newtonian gravity when the gravity is weak, and it does. General relativity is a superior theory that can explain a larger range of physical phenomena. In the century since general relativity was formulated, it has literally passed hundreds of observational tests. The tradition of mathematical theories and modelling as a way to understand the universe goes back to this time, and it continues strongly in modern cosmology.

A question that should come later, but I think it's probably very good to have it now, is what is for you the difference between the model and theory? Because here there seems to be quite a difference…

It's an important distinction. The underlying theory of the universe, the gravitational theory, is general relativity. That is a physical theory of gravity underlying everything we do in cosmology, because it's very well-affirmed through observation. The issue of modelling is distinct because general relativity encapsulates many possible space–times. You could describe anything in general relativity: black holes, the expanding universe, empty universes, you can describe all sorts of things, it's a very supple theory.

The modelling comes in because within the vast landscape that's possible within general relativity, there are particular world models. Cosmologists use the phrase 'world model' to describe the universe we live in, in other words, the actual universe as opposed to all other possible space–times and gravity situations that general relativity can accommodate. From the beginning of the exposition of general relativity, people realised it could describe any conceivable gravity situation — space–times that are highly curved or textured or wrapped in on themselves — so they homed in on particular world models, or models for our universe. Those more exotic situations might have applied at the time of the big bang, but the present day universe is large, cold, and the space–time is mostly flat. It's a rather gentle situation, without the extreme possibilities that general relativity allows. The basis of the model is that the expanding space–time is described by general relativity, and the physical objects of the universe like galaxies are simply markers of space–time. The main ingredient in the model is the mathematics to describe space–time as it expands. The formalism there is the Friedmann–Robertson–Walker metric. So in general relativity, the ingredient that you would call your model is the metric for the space–time, which is just a way of describing local and global geometry in that space–time.

So it could be, in a very naïve term, the movement of planets, or stars, or things like that with respect to each other?

Yes, the metric describes how galaxies move in the space–time, it will describe the curvature of space–time on small and large scales.

So, it's really a geometric metric?

It's truly geometric. In general relativity, the foundational theory, you can have disconnected space–time, you can have space–time that's tessellated, or sponge-like, or that has amazing geometry, all of this is possible. But luckily, in a sense (because the calculations would be very difficult), the space–time we live in is simpler. It curves gently, and we can describe the curvature, and so there's a basic metric. The interesting distinction — this is a very important aspect of the model — is that we are dealing with the model of the space–time, not a model of the objects it contains. And the objects in the space–time affect the shape of the space–time, so that is a major complication.

Would you say that the objects in the space–time are just what you can observe, with the tools — it's just your point of observation, but you try to model something much bigger than that. Does that make sense?

Yes, it's that, but also something else: it's the fact that in the geometric theory, its simplest and most obvious calculation is when all you have is the space–time, with nothing in it. In fact, the first solutions of general relativity were so-called empty universe solutions, because there was no matter or energy, just an expanding or contracting space–time. The complication is that the universe contains matter and energy. As soon as you add galaxies, and radiation, it affects the space–time and the metric of that space–time. Remember the metric is the mathematical description of the curvature of the space–time and the dynamics of that space–time, without anything else in the picture. When you apply it to a real universe that you can recognise and measure, there are a lot of galaxies. So in a conceptual sense, the galaxies don't really matter; for cosmologists, they only exist to mark space–time. It's a very strange situation. You have a theory to describe something that's invisible, since you can't see space itself, and time is a little puzzling, too. Cosmologists don't care about the properties of galaxies and the fact that each one has billions of habitable worlds. Galaxies are just markers to let you measure how the space–time is expanding, and tell you whether you had the right model in the first place.

So, I guess this is an interesting question that is coming, I'm modifying slightly the order: in many fields models are said to represent a target system — so you have something you want to model, you observe and you try to approximate as best you can, this is reality. So, how does this

apply here? It seems very complex in a way — because the reality is something you will never be able to observe, even if you have the best ever tool that you can imagine?

The relativity calculations are difficult. In a hundred years of calculation, only a handful of different metrics have been completely solved: the pristine black hole, the spinning black hole, and the one I mentioned, the empty universe model. If you go away from these simple or symmetric situations, it becomes really difficult to do relativity calculations; not many people are doing them, or are able to do them, so it's hard to make progress. A true calculation in relativity, to explain our universe, has to feed into it how much matter there is in the universe. The good news for the modelling is that the mass and energy in the universe don't affect the dynamics dramatically. It's a subtle, second-order effect. Cosmologists say that the metric is 'nearly' Friedmann–Robertson–Walker and so they can use perturbation theory to do the calculations.

But the effect of mass is important enough that it led, early in the development of the theory, to a prediction that the sum of the mass of galaxies and all the material contents of the universe should decelerate the expansion. If you have an expanding universe containing mass, the net sum of all that mass, of all those objects acting on each other, will retard the expansion or slow it down. Reversing that argument, if you could measure this deceleration, you could essentially weigh the universe, or measure the total amount of mass and energy in the universe, independent of any other type of observation. This was a major goal of observational cosmology in the 1950s and 1960s when the first 5m and 6m telescopes were built, telescopes large enough to count lots of galaxies and measure the expansion rate far away, so back in time. The universe has this attribute of 'lookback time,' where as you look out at space, you look back in time. Hubble was only able to measure the local expansion in the nearby universe, which is the contemporaneous universe, because these galaxies are maybe tens of millions of light-years away.

So, they are too close to see anything?

Right. The universe is roughly 14 billion years old, and these nearby galaxies are in the most recent 1% of the history, so they're recent in cosmic terms. By the 1960s, you could measure objects that are one or two billion light years away, which is a significant fraction of the 14-billion-year age of the universe, and in the theory you're predicting that you should see this deceleration. So, whatever local expansion rate you measure, as you look to these more distant galaxies, their expansion rate should be faster because the universe has decelerated since then. The hypothesis is clear, but the observations are very difficult. Measuring the local expansion rate as Hubble had done took the Hubble Space Telescope a huge amount of time and wasn't properly done until the late 1990s. The complication of measuring more distant galaxies, where you have to make assumptions about the physical nature of the galaxies and their brightness and whether they're the same as galaxies nearby, this has all been very hard. In cosmology, you have this unavoidable problem that you're looking at the nearby 'now' and the distant 'long ago,' and you cannot ever bridge those two. You cannot prove that a galaxy that you see five billion light years away is the same as the galaxy now because you're observing it then, and there's no way round that. The finite speed of light and light-travel time present you with a fundamental obstacle.

So, would you say it's a lot of conjectures and then some observations are possible to make, and the conjecture then is verified or there is a contrary example and that's how the cosmology science evolves by massive steps? Does that make sense?

In the modern development of the subject, advances have often been limited by the difficulty of the observations and their interpretation. The theory of general relativity is very mature; the calculations are challenging, but they're not impossible, and now you can do them numerically with computers. In the past 20 years, numerical relativity has become its own subject and computers help in dealing with the complexity. But in terms of foundations of the subject, another issue is the assumptions that are made or that have to be made in cosmology.

Yes, that's quite important.

I should talk about assumptions. Relativity is a self-contained formalism, that's either valid or not and you can decide by experiments. But when you actually make the model for the universe, then you start to have to make assumptions. One of the assumptions is that the space–time is smooth and connected. There's a general mathematical way of describing it, but basically that means that the space–time does not have any discontinuities or tears, and it also means it's not topologically complex, like a pretzel or a knot. In general relativity, you can describe arbitrarily convoluted space–times, and people love that stuff, but for doing practical cosmology, you assume the space–time is simple and simply connected. A little boring. A related assumption is that the curvature locally, and even globally, is modest. These are assumptions that make the calculations tractable, and it's gratifying (and perhaps a little lucky) that they've been verified. Observations of the microwave background radiation from the early 1990s were able to demonstrate that the space–time that we live in is flat to within 1% or 2%. There's also no sign within the volume available to observations that space–time is not smooth and continuous.

However, some cosmologists are happy to note that we are limited in our observation by the space–time available to us, in the sense that the observable universe is not bounded in space, but in time. All we can observe with telescopes is the region where light has had time to reach us in 13.8 billion years. Beyond that, the space–time plausibly continues, and it could have different curvature. So people like to speculate about wrapped space–times and other things that happen beyond the places we can see with our telescopes. But the parts we see with our telescopes seem to behave simply, so this is pure speculation.

Two other connected assumptions are absolutely foundational and were in play from the beginning. Together they form the cosmological principle, an important premise about symmetry that underlies the metric used to describe space–time. The cosmological principle is the assumption of homogeneity and isotropy for the universe. Isotropy is the one that we have tested and we believe that the universe is isotropic. That essentially means that if we look out from the Earth in any direction as far as we can, measure the population of galaxies, the expansion rate, the curvature if we can do it, the properties going back in time, we see the same thing anywhere we look. Isotropy is testable. With the Hubble Space Telescope, we can look at extremely faint galaxies and if we check, we find the same numbers and populations in another part of the sky. With the microwave background radiation, we see speckles or fluctuations, and they have the same properties all around the sky. So isotropy is affirmed as an assumption of cosmology very well, actually to a precision of better than a percent. Homogeneity is much harder to test, because it says that our local region of the universe is not bizarre or atypical of the whole observable universe. We can't test that — we're stuck in our local space–time and we observe distant regions as they were in the past. That means we're comparing the recent nearby universe with the distant younger universe; since the universe evolves, the situations are not directly comparable.

It's also very difficult to test whether or not the laws of physics we measure in the lab apply across the universe. A whole sub-field of physics and cosmology looks at the basic laws of physics that describe the universe — what if physical constants change or decay? If they do, cosmology would be profoundly affected. That's another homogeneity issue. So far, there is no evidence that the laws of physics change across time or space, but the experiments are very difficult and limited.

The assumption implied by the cosmological principle is built into the model, because the Friedmann–Robertson–Walker metric in an expanding space–time is based on homogeneity and isotropy. If that assumption is wrong, those calculations are wrong. I'll give an example of an aspect of homogeneity that we *have* tested which many people are not aware of: the relative amounts of matter and anti-matter. The universe is overwhelmingly made of matter. Anti-matter, which is produced fleetingly in labs given sufficient energy, is extremely rare on Earth. But people have wondered whether that is true on a cosmic scale? Microscopic physics is highly symmetric, you can produce particles and anti-particles equally out of pure radiation, and it happens all the time in the vacuum of space. Maybe our matter-dominated part of the universe is atypical, and there are anti-matter sectors out there. This is subject to observation, because even if you posit anti-stars or anti-galaxies or even larger anti-matter regions, there are sufficient interfaces even in the vacuum of space that would generate gamma rays. Those gamma rays would be detectable by satellites and they haven't been observed. As a result, there are very strong limits on anti-matter in the universe. So our universe is homogeneous, in the sense that it's made of stuff instead of anti-stuff, everywhere we can look.

Let me address another foundational issue: the validity of general relativity to describe the universe as a whole. So far, relativity has passed its tests with flying colours, and one test that it has passed fantastically well is the bending or deflection of light due to a massive object. This phenomenon is called gravitational lensing. If lensing is weak, the image of a distant object is distorted or stretched out, and if lensing is strong, multiple images of a distant object can be formed. This key prediction of general relativity was first observed in the late 1960s, and by now we've seen hundreds of thousands of examples. Each of these is a little gravitational optics experiment where we can see light bending through large volumes of space because of the mass and energy in that region of space. Even though the universe is globally flat, there are a lot of interesting corrugations and ripples on smaller scales. General relativity describes these deviations from flat space–time very well.

This is clear. So, just going back, because we spoke a lot about quantitative tools and mathematics and its role; and we spoke a lot also about observation, possible observation and possible testing, would you say mathematics, some observation test, if possible tools — are these the key constituents of modelling in cosmology?

Yes. I'm choosing to talk about cosmology as the study of the universe as a single entity. My research is more the evolution of galaxies and supermassive black holes within an expanding universe. Now, I'm talking about people who only care about the behaviour of the entire universe.

There's a question a bit later about simulation and the use of computers, is it appropriate to speak about it now? What is the role of computers — I guess they play quite a huge role in simulation, maybe to solve some system of complex equations, or…?

Computers have played a really huge role. In the 20 years or so in which they've been powerful enough to really make an impact, they've been important in two areas. One is in numerical modelling of general relativity, making approximate solutions to calculations that are brutally difficult. We've had the formalism of the theory for a hundred years, and the number of experts is not that big, but they have only solved three different metrics completely. Full, analytic solutions of gravity situations in general relativity are extremely difficult. Computers can help with that situation, and they've been transformational in being able to calculate scenarios like the immediate environment of black holes or the phenomenon of gravity waves.

The second area of astrophysics where general relativity applies and where computers have played a central role is the description of large regions of the expanding universe. In this type of cosmological simulation, the model is an approximation because the space–time is not empty, it's filled with galaxies and radiation. The computer lets you do that approximation accurately with a computation. These calculations have been done with billions of 'particles' standing in for galaxies, and for example, they've shown how the universe could evolve from smooth initial conditions to a highly structured situation of galaxies and clusters and voids. However, these simulations make some theorists a little nervous, because they say that as soon as you just trust the computer, it's only as good as the physics you put in. We think we know what the physics is, but if the simulation produces some surprising or strange result, how should we trust it? There are esoteric and occasionally acrimonious debates between pure relativists who calculate in the general relativity formalism, and those who simulate gravity situations where general relativity is an ingredient and there are other ingredients and all sorts of computational trade-offs and approximations. The use of computers to simulate the universe is productive, as long as the simulations are tethered in tested physics and checked against real observations.

Would you say that simulation somehow is a complement to observation, and can replace observations that are not possible? Could it be seen this way or also as an additional tool in parallel?

Yes. Cosmology has evolved from being a two-cornered situation where you have foundational theory and you have observations made with better and better tools, and you're checking the theory against the observations. The theory suggests interesting observations, or observations that might disprove the theory. With the big bang model, the testing now has moved back to the first pico-second after the origin of the universe and the physics is based on lab or accelerator experiments. Now, the situation is a triangle and linkages in each of the directions are very strong. The theory and the observations are still critical, but the simulation–observation relationship is important because simulations produce an accurate representation of the universe in a computer, and they suggest a whole range of phenomena that you might observe that the simple theory wouldn't predict, or wouldn't allow you to calculate. In principle, simulations let us test the theory at a deeper level and with more varied diagnostics. This large and growing field is always subject to a big caveat — don't trust the simulations too much. Many researchers in my field spend their time 'observing' simulations.

So, you even have the observation of the simulation...

I'm being tongue in cheek now, but science could turn into solipsism since it's a lot easier to study a simulation than to go to a telescope where it might be cloudy. I can observe a simulation that's sitting on my computer. I can draw 10,000 galaxies from a simulated chunk of the universe and see what their properties are and write a paper about it. People do this all the time. So the triangle has become quite interesting. Computation power increases with Moore's law and it becomes so much better that it's increasing the type of problems you can address and the subtlety of the observations you can match with your simulation. Remember that true general relativity calculations are hard. I had a colleague who was a relativist and we used to teach cosmology to graduate students in the program here. I did the observational side and he did the theory side. (He's a Jesuit priest, so he lives a double life; he has parish where he's a priest and he's a PhD scientist). Once I asked him 'How many people understand general relativity?' I know how many people have taken a general relativity course; I have and several thousands of other physicists and astronomers have as well. I didn't mean that; I said 'How many people in the world understand general relativity here in the gut?' He said 'Twenty.' Those are the high priests (and a few priestesses) of the profession who carry the flame of Einstein. Meanwhile, hundreds of people do simulations so the gravity priesthood has expanded thanks to computers.

In some fields, there is a particular order — you do for instance observations, model, and then simulation, and then you go back to the observation. So you have a sort of logic between these three entities. In cosmology, would you say there is a logic or can it go in this triangle, can it go in any direction?

It goes in all the directions. You have legitimate research projects and interesting research questions among any pair among observations, theory, and simulations. But I would say that science is rooted in observations so they are essential. I'm not the kind of Platonist who believes we can understand the universe by pure thought.

In the modelling process, let's say you focus on black holes for instance, for the modelling itself would you have a certain order that is more logical to model? Or will it make sense to simulate before looking at certain mathematical formulations of something, or will the simulation be done afterwards or during the mathematical formulation?

It's not always a strict order, because sometimes you alter the model as you think of the observations you want to make. I'll give an example: one of the ingredients of the expanding universe model, which has to be part of every simulation, is dark matter. Dark matter is not understood in a physical sense right now, though we hope in a few years that it might be discovered as a new fundamental particle. Dark matter is an ingredient in all cosmological models, but it's an X-factor, it's a mysterious substance, so it's inserted in a phenomenological way rather than a physical way. Astronomers have eliminated many of the possible candidates for dark matter. It can't be burned out stars, or black holes, or rocks, or dust particles in space; it's almost certainly a new subatomic particle. But we don't know its basic physical properties.

So dark matter is an ingredient in a model. It exceeds by a factor of six all the protons, neutrons, and electrons in the universe. We have measurements of the amount of dark matter on different scales from observation, and they feed back into how to put dark matter into the model. Given our ignorance about the physical nature of dark matter, simulators and theorists quite legitimately say we're going to try all sorts of dark matter in our model. We're going to experiment with dark matter that interacts with itself or doesn't, with dark matter that's hot (in the jargon) or cold, which just means whether the fundamental particle was relativistic or not early in the universe. They will hypothesise all sorts of dark matter given that we don't know what it is; they are experimenting with ideas. Observers don't necessarily care about that type of speculation because putting hypothetical forms of dark matter into the computer simulation may or may not lead to something they can observe. However, the modellers and theorists consider this totally legitimate work in its own right. It's a matter of taste as a scientist. There are theorists who are much more oriented towards producing a prediction that can be tested and then work tightly with observers to do that, and then there are some that aren't.

The power of computers has enabled a new form of cosmology. It's counterfactual science: thinking of ways the universe could be, but isn't, and trying to learn from that. You can do so many simulations in the computer, you can alter your assumptions or vary the physical properties of the dark matter or the ingredients you put into the model and run huge numbers of simulations. We're far from the days where it was a challenge to run one simulation, so you'd make your best estimate of the ingredients and the initial conditions and try to produce something roughly approximating the universe that you observe with telescopes nearby. We've gone way past that. Now, we use this fantastic CPU-power to do hypothetical simulations, to vary parameters, and to alter assumptions. As a result, progress in cosmology is accelerating and it's very exciting.

OK, I think this will be quite interesting when we go further down to the question of uncertainty, and things like that.

What would you say is the aim or use of the model? For example, some models are for learning or exploration or exploitation. Here, it seems it's about understanding the universe or learning about the universe. What would you say is the ultimate objective of a model?

It is all of those. You are learning and exploring, but you're also testing — the core of the science is constructing the model based on a physical theory and seeing how well it matches the observations. Most cosmologists will say it's still an empirical science, still observationally-driven, that if we move too far from being connected to observations the subject becomes unhealthy. At the cutting edge of cosmology these days, people talk about the multiverse concept, other space–times emerging out of the big bang, string theory, and so on. That's an exciting, speculative froth of the subject that is still not truly testable, so hard-nosed people say remember the scientific method. They'll say fine, you can speculate about all that, but it's really philosophy, it's metaphysics at some level, until you can test it. I think the core of the subject is still very observationally tethered and oriented, and that's probably a healthy situation.

The best example is the microwave background observation. When it was discovered in 1964, it was a dramatic affirmation that the hot big bang was the right model of the universe, because it blew the other models out of the water. How else do you explain why every cubic centimetre of space has a hundred million microwave photons in it? The big bang perfectly explains the observation as the relic radiation left over from the big bang after the universe has expanded and cooled. Right from the start, the microwave background radiation was a lynchpin of why we believe the hot big bang model. Now, fast-forward through 50 years of fantastic innovation, and we've progressed from 4 meter telescopes to 10 meter telescopes, and we have the Hubble Space Telescope and sophisticated satellites. When the microwave background was first measured, all we could say was that it had a temperature of just under 3 Kelvin and the same intensity in every direction. Now, we have exquisite data — six or seven orders of magnitude better than it was in 1964. We measure variations to one part in a hundred thousand, and we're testing gravity theory and parameters in the cosmological model at that level. It's called precision cosmology. As a result, we have precise measurements of the age and expansion rate of the universe, the curvature of space, and the proportions of normal matter, dark matter, and dark energy. But this precision coexists with ignorance about the physical nature of dark matter and dark energy, ingredients that account for 95% of the universe. It's a strange situation.

Coming back to a question which we didn't discuss earlier, which I think seems quite interesting for cosmology: what is the role of language in

modelling, and more qualitative aspects? Because you mentioned philosophy earlier — I mean the link between philosophy and cosmology, and we are touching on something quite fundamental in a way, so I guess language has to be quite important.

As a scientist, I'm intrigued by the role of language. My first discipline of course was physics and in physics the language is really important: people talk about particles and waves and fields and, really, what does that mean? Quantum physics tells us that these concrete words can be misleading because nature is supple and microscopic phenomena are both wave-like and particle-like. Energy and matter are evanescent and shape-shifting and there is indeterminacy at the heart of the quantum theory. So the words are at worst crutches, and at best, imperfect models. In cosmology, it's the same thing. Most practitioners would say that the theory is fundamentally mathematical. The theoretical core is general relativity, that's a mathematical theory, there's no ambiguity — you can either solve the equations or you can't, it's just a technical thing. The substrate of work in cosmology is not ambiguous and not open to interpretation. Words aren't essential; mathematics is self-contained and self-referential and it may even be universal. If you go to an international meeting of theorists, their papers may be in Russian or Spanish or Mandarin, but from the equations you still know what they're doing. In these mathematical arenas, the language doesn't matter at all.

When it comes to applying the theory to the universe, then the language starts to be important, but I think the core terminology, the jargon words, the technical terms, are all clearly understood. When cosmologists talk about 'space–time,' everyone knows what that means, when they talk about a 'metric,' that's not a colloquial word at all, that's a precise technical term. With other words like 'galaxy' it starts to get a little more open to interpretation, but we still understand that's a self-gravitating system of stars. Cosmology is not for the most part language-dependent; there's a common set of terms that are very much rooted in the theory and the mathematics and so they're not open to interpretation. So cosmologists get to avoid some of the philosophical issues that arise from the language of physics, except in situations where micro-physics is important. But I don't see many situations where language becomes an obstacle or is ambiguous or difficult to interpret.

How important is notation?

Every subject has its own notation, and cosmology is no exception. On the theoretical side, general relativity is expressed mathematically. But there are at least four different formalisms of notation for the theory, and textbooks that use each one, which causes headaches since the goal of notation is to have a common mode of expression. The core of relativity is a set of 10 coupled, second-order, partial differential equations. That's quite cumbersome to deal with, so there is notation that compresses the formalism into one equation, a shorthand that assumes understanding of tensors. The single equation relates the curvature of space–time to the density of mass–energy. Observational cosmology has its own notation, notably the key cosmological parameters: H, the Hubble parameter, Ω, the density of normal and dark matter, Λ, the cosmological constant, and k, the curvature of space–time. In the full cosmological model, there are another dozen or so parameters. Cosmology is a small enough field that the same notation is shared and understood by all practitioners.

And the last question about modelling: what is for you a good model?

A good model for me is one that is rooted in verified physics, and is tightly tethered to or constrained by state-of-the-art observations. It's a mathematical description that can be held up against the real universe and be verified, or modified, or potentially refuted. That's the gold standard and that's pretty much what everyone aims for. There is also the hypothetical, counterfactual aspect of cosmology, where the model can be varied widely and normal assumptions can be relaxed or discarded entirely. There have been times in the history of cosmology when the coupling of theory and observation was not so tight. There were times when the observations were just so difficult you couldn't get good data, so you couldn't test the theory. There have been times when theorists worked so esoterically on things that you couldn't observe that observers said you're wasting your time, why don't you explore an idea that can be tested?

Yes, that makes sense. So, now we have two sets of questions. In certain fields, I'm not sure if it applies to cosmology, there is a huge difference between risk and uncertainty, so that's why I have this differentiation

here. How would you define uncertainty, and how does the model help you understand uncertainty in cosmology?

Now we are into an important language issue, because the word uncertainty is important. I first go back to physics, because Heisenberg's uncertainty principle is at the heart of the measurement and understanding of the microscopic world. Many physicists object to this particular terminology because they consider that Heisenberg's principle is not about uncertainty in the way we understand that colloquially. The uncertainty principle is really about imprecision — the floor on the knowledge you can have about certain physical properties, whether it's mass and momentum, or time and energy. As far as we know, Heisenberg's principle describes a limit to knowing, regardless of the nature or quality of the observations.

In cosmology, the word uncertainty is used in several different contexts. Let's start with observations. Data are never perfect, so every observation has an error or uncertainty attached. Hopefully, it is just understood in terms of photon statistics or the design of the detector, but there can be systematic errors that skew a result. In the history of astronomy, systematic errors have been hard to diagnose and have led to some famous blunders. When observations are compared to a model, the model is accepted or rejected based on a 'goodness of fit,' which is usually applied in a chi-squared or maximum-likelihood formalism. If the observational uncertainty is too large, there is poor discrimination between models. Progress is made by designing more precise instruments and detectors and gathering more light or radiation and so driving down the observational uncertainty.

In terms of underlying theory, there is uncertainty about the limits of applicability. Despite me saying that general relativity has passed all the tests, the truth is that it's not necessarily the ultimate or final theory of gravity. In the work of Stephen Hawking and others, it's clear that general relativity fails when it tries to describe black holes, because it produces the prediction of a singularity, which is un-physical. Singularities mean you can't calculate, they mean your equations are breaking down. A different type of singularity arises in cosmology when we push the big bang model back to the origin. The initial state has infinite density and temperature so it's also a singularity, and it's pointing to the fact that we don't have a theory to unify gravity with quantum physics. Despite its successes, general relativity is not the ultimate theory of gravity. Luckily, most aspects of cosmology involve relatively weak gravity and fairly simple, flat space–time where general relativity works very well.

Another level of uncertainty would be of all the other assumptions you make, some of which you can and can't test, such as the cosmological principle. If the universe has large-scale variations in density and is not homogeneous, we can't test that. I could even raise causality, the assumption that events have causes. Science as a whole subject would fall apart if causality were not true, so I'm not even going to question that. We're just talking about these assumptions, some of which you can and can't test, or you can test well or not so well. That would be the next layer of uncertainty.

The next layer would be the ingredients that you put into your model when you go from the theory to the model, and there the huge uncertainties are dark energy and dark matter — the two ingredients that constitute 95% of the universe and which control the cosmic expansion and we don't know what they are! Given our uncertainty, some wild ideas are on the table, especially for dark energy. For dark matter, we may be closing in on a candidate, subatomic particle, but current searches might fail. Dark matter and dark energy are each huge challenges for the standard model of physics. In cosmology, you just operationalise through your uncertainty; you just say well I don't know what dark matter is, but I know how to measure it — I know how to weigh this part of the universe, or see how much dark matter there is in this galaxy or in this cluster. If you do that, it's just an ingredient in the model and you don't have to know the exact nature of the particle.

Yet another type of uncertainty arises in a variety of messy ways, and it's the sense that the model is not quite good enough. I don't mean in the general relativity sense, I mean how the ingredients of the universe interact within the space–time, so how space–time curves because of galaxies and radiation. Galaxies evolve and convert gas into stars, and you can't measure that very well. They also interact with their environment and each other. Cosmologists have a word for this: gastro-physics. Instead of thinking of galaxies as pristine little markers of space–time, they're not pristine at all — they're big messy systems that have gas going into them, and gas coming out of them. They interact and merge and share matter, and there's a lot of hot gas involved, so it involves shock heating and complicated physics that's not very well put into these models or the simulations, it's very hard to do that in the simulations. The messiness of galaxies is another whole level of uncertainty, and even if you thought you only needed them to mark space–time, you can't pretend they're little points. The complexities of galaxies need to be understood.

Let me add another one. It's the uncertainty attached to the limitations of the simulations. When you simulate a universe, it's called an N-body calculation because you're calculating the gravitational force of N particles on each other. The biggest simulation might have a hundred billion particles, so these hundred billion particles interact in an expanding space–time and that's all you can calculate with your biggest super-computer in a reasonable amount of time. There are choices and trade-offs to make: you could make each of your hundred billion particles one galaxy, and your simulation would be of the entire visible universe, or you could make each of your particles one star, and the simulation would be a single large galaxy, or you could go anywhere in between. Simulators have to parse the available computational power into the size of the space–time being modelled and resolution or mass corresponding to each particle. After a set of calculations, all the particles move simultaneously by gravity, so there is a step size in time, then the calculations are repeated, and so on. Since the gravity force declines quickly with distance, you don't have to calculate the gravity of every particle on every other particle. Some computational short cuts can be taken. Many computational strategies involve approximations of the physics, so they amount to uncertainty.

Let me talk about a final type of uncertainty — another foundational one. There's an issue in cosmology called cosmic variance. It amounts to the fact that you only have one universe to observe, and that becomes an uncertainty. You're trying to observe the entire universe, which is by definition a unique object. In the standard big bang, there are regions of space–time that we can't access with telescopes, because the universe was expanding faster than light for the first third of its existence. So, if we have a hundred billion galaxies in the observable space–time, we know that there's more space–time out there and probably lots more galaxies, and the fraction of space–time we have been able to observe might be a very small fraction. This means that the physical universe, all space and time, is a lot bigger than the observable universe. So, there is a sampling issue, it's almost like the cosmological principle all over again. The issue is not just whether our local region is typical of the observable universe, but whether or not the observable universe is typical of other regions of space–time that we can never access with our telescopes.

This leads to speculation about the earliest phase of the universe, the big bang itself. The big bang has been taken back to the arena where the universe was smaller than an atom — it even makes sense to talk about a physical state where matter that would eventually become a hundred billion galaxies was concentrated in a region smaller than a proton. So cosmologists posit that the big bang was a quantum event, and the universe emerged from a quantum fluctuation. Continuing the speculation, a substrate reality that led to our universe might have co-existed with other quantum fluctuations, some of which might have inflated to a large size. That's the multiverse concept. In a multiverse scenario, our entire space–time is part of an ensemble, an ensemble that we don't understand very well. This leads to the natural question of whether or not our universe is typical, and an anthropic argument that unless you happen to live in a universe able to make carbon and stars and water, we would not be here to talk about it. All of those other universes with properties inhospitable to biology do not generate observers like us. So, the fact that we only have one universe to observe is actually quite profound.

It seems you have these levels of uncertainty in the modelling phase, if you like; so, this is part of the modelling, but it is not the model trying to explain the source of the uncertainty, which is a bit outside of the model, if that makes sense?

Right. This is just the model. As I talked about earlier, you have the uncertainty of the observations, of the data, which has its own whole set of limitations. Some of them are limitations of space and time, connected to our situation in the universe, such as which galaxies you can actually observe with a telescope because the light has been able to reach us? Then there's the question of which galaxies you can observe where you can gather enough photons in a reasonable amount of time to measure anything interesting or relevant? Then there's the question of how can you get a big enough sample of galaxies with high enough quality data to compare meaningfully to a simulation or a theory? These are levels of uncertainty associated with the data, or the observations.

That makes sense. And do you have a notion equivalent to risk? Do you make a difference between uncertainty and risk, or does it not really apply?

Risk is not a notion that is encountered very often in cosmology. Risk can apply in astronomy when it comes to space debris and the probability of an impact on the Earth, or risks to humanity from rare but high energy cosmic events, like supernovae or gamma ray bursts. These are risks associated with our immediate environment. Risk quickly becomes very abstract when thinking about cosmic scales of billions of years or billions of light years.

This is quite closely related to climate I think, so that's interesting. So, is there this notion of risk, or not really?

Not really. I would just be facetious because cosmology is pretty esoteric, nobody's life depends on whether we get our answers right or not, so in that sense, what is the risk of being wrong? Not really very much to anyone. So I don't know if cosmologists talk about risk; they talk about the risks to their own careers of going on a wrong path, or spending time working on something that doesn't pan out or a theory that doesn't pan out. An example that got into the news a few years ago was if dark energy has certain properties, then there is some existential risk to the universe, and that's something that nobody knows is going to happen on a human time scale or a geological time scale or what kind of a time scale. But no, I would say in cosmology, we don't talk about risk very much.

It is very interesting because you give an answer that is very similar to that of some other experts: risk is somehow a risk to others or society.

I would add one thought to that about risk. It's not as profound as a risk to the culture, but I think that cosmology still has some fundamental issues to resolve, so there is a significant risk if we are going down the wrong path with our gravity theory or our world model. There's a little asterisk in everyone's head asking this: what if we are fundamentally mistaken about some core premise of cosmology? As an example, you can explain away dark matter in some alternate theories of gravity. If gravity is not a perfect inverse square law, you only have to modify the nature of the force law of gravity by very small amounts, a few percent, and you don't need dark matter, just the normal stuff. But to explain away dark matter, you've given up something profound, because Newtonian gravity and general relativity are self-consistent in a mathematically profound way. So, if you just mess gravity up enough to explain away dark matter, you've ruined the beauty, elegance, and symmetry of the theory. But we know that general relativity is not the final answer on gravity, because it doesn't accommodate quantum effects, and theorists will acknowledge there's a risk that we're not on the right track. The difficulty is that the bar is set very high for an alternative theory of gravity, because it has to explain everything we've already explained and then additional phenomena. String theory suffers from a similar problem — there's risk attached to working on string theory, because as beautiful and elegant as the mathematics might be, and as much as you have hints that something like that may be the fundamental nature of matter, it could be completely wrong. String theory could be a huge dead end, just a diversion.

So, in this sense, would you say the risk is the ultimate uncertainty, the maximum level of uncertainty in a way?

Yes, it's really the risk of being badly wrong. Of course, that's part of the thrill of science. Nobody knows the answers to certain simple and fundamental questions about nature. To make progress, you have to be bold and be willing to fail.

And the last question is more a personal question. In your field, what is for you the result or the theory or the experience which has had the most significant impact, and why? It's not necessarily a result you discovered yourself, it can be something more general.

In the field that I work on, a key insight of the past few decades has been the fact that the universe is extremely good at making black holes of all sizes. Black holes were predicted as an outcome of intense gravity in the core of a massive star when it dies, and we have detected several dozen good examples of black holes a few times the mass of the Sun. The black hole itself is invisible, but they are in binary systems with a star that is visible and with material heating up and glowing in X-rays as it falls into the black hole. More recently, we have discovered that every galaxy has a massive black hole at its core. I work on black holes which range from 10 million to 10 billion times the mass of the Sun, so nature is somehow able to make black holes all the way from a few times the mass of the Sun up to 10 billion times the mass of the Sun. While every galaxy seems to have a black hole, most of them are innocuous and were unnoticed for a long time because they're not active. When a black hole isn't consuming stars and gas, it is quite dark. The ubiquity of black holes in the universe over such a vast range of scales is amazing to me.

So you have a good dataset!

Yes. Another thing, just as an aside because people think black holes are so exotic and so strange, the black holes I work on are not as strange as you might think. The big ones, ranging from the Milky Way at four million solar masses to the nearby elliptical galaxy M87 at three billion solar masses, actually have a density inside the event horizon that is less than water. The way the scaling of the mass and the radius of a black hole goes, the density actually goes down as the black hole gets bigger. It's generally thought you would want to avoid a black hole because the gravity stretching force would kill you as you fell in. But for these massive black holes, you get around the problem of being ripped apart, because the total gravity is strong, but the stretching force is mild. So you could go into these black holes and figure out what happens inside them. It's not feasible, they're much too far away, but it's interesting to me.

I think that was brilliant — it was really interesting. I hope my questions weren't too naïve because you are really far from my field!

When you sent the questions, I immediately realised how well they fit. I appreciate that you've been able to ask questions that apply across these diverse fields, because I know interdisciplinary work is hard. But it was obvious as soon as you raised the issue of modelling — yes, I can talk about that, that would fit cosmology and I can imagine how it fits the other people you're talking to. So, I congratulate you on very well-defined questions that challenged me to think about my subject deeply.

Chapter 10
A Conversation with Gerd Folkers

Can you describe your field briefly?

My field is molecular modelling. The background is pharmaceutical sciences. So the aim would be to create, design, construct, whatever that means (we can come to that later) a new therapeutic. Since, in the classical world, therapeutics are, to 80%, molecular-defined entities, it would translate into creating a new molecule, by playing with existing ones.

Present a paradigmatic example of a model in your field, describing it in terms that are accessible to non-experts.

A paradigmatic example is... I do not know whether this is already a model — that's interesting — or whether it's a metaphor. It is a famous saying, or a famous illustration of Emil Fischer, from around 1904, something like that, who was among the first to apply biochemical reactions in organic synthesis. So, to build up a certain molecular structure, he did not use classical chemistry, he used an enzyme from nature, from a bacterium, and let this enzyme do the special synthesis; and then removed it at the end, and it went on. This was exciting at those times; and he had to explain this in a lecture to his colleagues from the *Gesellschaft Deutscher Chemiker*, a very conservative [society]. And so he had difficulties explaining how these two entities — the enzyme and the molecule to be built up — would interact, because the molecule's structure was spectacular, since it showed so-called chirality, handedness; a difference between left and right, hence you had to do it in the right way. And the enzymes, the proteins, could do that, but it's very tedious to do that by classical chemistry. And so he coined the word that these two entities interact like a lock and a key. So, you have a perfect cooperation, or mutual recognition, of the key and the lock; and that means the entity number one should fit, at least geometrically, the entity number two, to have a perfect reaction.

So, would you say that this conceptual representation of the lock, which would be like the model, came after experiment? We'll come back to this later, but…

Right. We'll talk about targets later.

So, what you would describe as a model in this case, is the last phase — somehow the abstract representation of the lock and the key. Am I correct?

Yes, you are correct. Therefore, the question as to whether this is a model or a metaphor. I mean, it is a model insofar as it is the illustration of the theory.

We'll come back to theory later as well!

So, the illustrated theory that he hit upon, the interaction of the molecule with the protein, and so he coined this idea of lock and key, and this is still along the lines that we think on today — even though this is 120 years ago. And still, today, we think that the so-called receptors — the proteins inside our bodies that recognise therapeutic molecules — are the locks, and that the molecules are the keys; they have to interact in a certain way, to move somehow the lock and then to have an effect.

With the help of this example, could you explain why a model is needed, and more precisely can you describe what the model is? So, here we are really about understanding maybe the various steps in the modelling.

OK, why is this so important and, in my opinion, what was the breakthrough? Up to then, you had more or less, to coin it a little bit cynically, strange ideas why a therapeutic entity with an active ingredient would work, in humans. This was very strange. So, of course, humans had their first experience in the biological activity of molecular constituents by being poisoned. And they did believe they could understand this, for completely the wrong reasons — because most of them were spiritual. So, this very famous example of African tribes who prepared poisoned arrows, in a way that it was a 24-hour ritual, where they cooked poisonous snakes and everything that they knew was life-threatening — scorpions, whatever — all these poisons of these animals are destroyed by cooking, because they're proteins (like egg, if you cook an egg); then at the end, they put some leaves from trees in the soup, and the trees contained a molecular structure that is *Strophanthin*, the leaves of the *Strophanthus* plant. And this *Strophanthin* is absolutely poisonous, because it makes your heart stop. So, the only remainders after 24 hours of cooking was the stable molecule that they extracted from the leaves, but they did not know that. But it worked extremely well, so all their enemies were dead. So, the reason why something should work, to transfer the poison of the animal into something extracted in principle was right, but they had no clue about what was happening. Then in the next stage they had something that is called in German *Signaturenlehre*. *Signaturenlehre* means hermeneutics of Nature, or exegesis of Nature, of which the most prominent example is bio-Viagra. So the normal Viagra we know is a small molecule, it increases blood flow, etc....; the "bio-Viagra" is a [pulverised horn of the rhino]. So the *Signaturenlehre*, it's self-explaining, it's clear. So if you take the powder and then put it on some tea, it has the same effect. Of course, it does not have the same effect, but you can imagine that it has the same effect, so it's a giant placebo effect. And so from the *Signaturenlehre* the most famous example is aspirin. Salicylic acid, which is a derivative form of the aspirin or vice versa, is contained in the bark of the willow. And the willow tree always stands with its roots near water, like a brook or little river. So people, it was a British reverend in the 18th century, assumed that standing with the feet in water for more than several years would cause permanent infections, flus. But because the willows did not die, they should have an inherent principle to combat this [sort of] infection. Therefore, he recommended if you have an infection, a flu, to take the bark, prepare a tea, and drink it. It was completely right but for the wrong reasons. He extracted salicylic acid, which today we know as aspirin.

That's really interesting.

So again, it was *Signaturenlehre,* a hermeneutic interpretation of what they saw in Nature. So, that was for some 1000 years, there was this deep prominent desire for new remedies.

So, would you say, given what was available at that time in terms of tools, techniques, thinking processes, do you think that these principles as you call them could be referred to as models in a way?

This is a very difficult question. In fact, funnily enough I have discussed this very often with my wife's director who is a philosopher of theology. So what does it mean, hermeneutics, and exegesis? What does it mean, the interpretation of what you see, what you read — is this a model? I don't know, I'm not sure. In principle, you could say the willow is a model, and on the other hand, of course it's not a model in structure but it's a model in fighting some disease, which is based on a completely wrong assumption, which may be typical of models! On a wrong assumption that there is a disease; but for willows there *is* no disease, but still they have something.

That's really interesting, because then we go really to the thing of what makes a model a model.

Yes, what makes a model a model?

Because somehow the model exists because there is somebody thinking about it, right; and this British reverend maybe was doing modelling in a way.

Yes, he was doing modelling in a way, because in his mind he had a logical process which was derived or used from a theory that he had about Nature. There was a theory in the background, also kind of an evolutionary theory — god has prepared all the things in a way that they could help themselves and so on and so on — that was the basic theory, very probably out of the Bible. In his framing of the world was this completely logical step to assume that things like that would happen; and for the rest of the world, they simply had not yet detected it. It was also positivist to say this is true for willows, so maybe liverworts have different jobs, little yellow flowers, are very famous for being good for the liver because if you have liver problems, you are yellow, and so on and so on. There are very many of these things still followed in the name of the popular botanics today; so in German we say *Leberblümchen*, or yellow flowers for the liver; we say *Augentrost*, which means comfort for the eyes, and so on.

We will come back to this maybe when we discuss model and theory, but just because it seems quite appropriate here, would you say that you have a general context, which will depend on historical and where we stand in terms of philosophical, religious, any type of status? And then you would have an underlying theory, and then from there individuals can develop models? Something like this?

Yes…

And when there is an evolution in the context, obviously everything shifts.

Sounds good. And before that, again, everything was based on a theory that came from the Greeks, or, no, it was from the philosophical half-moon, so it came from Asia to the Greeks; so from India, probably from China too, there was the theory of the four elements. It's very well known — the Japanese had six elements, the Indians I think had five, the Greeks had four because they thought four was a magic number, and so on — you know better the mathematics. The Greeks loved four, and so they reduced to four elements. And then they explained everything with these quadrants, with the four elements, and then you had these phlegmatic types, these humours, and then they addressed certain remedies to certain types; which again is a model in my…it again has some truth — there are parts that are not so wrong, and…it's interesting.

Yes, it's very interesting to put the model into a context — it's quite new here in terms of my interviews.

So then the different stages are important.

Yes. So let's say we are now, today — what would be the key stages in modelling at work?

We have all these gigantic molecular biology tools of today, and this has changed a lot. It has turned upside down the parameters in research, especially in pharmaceutical sciences. Because now we can start from the genetic description of a disease, to extract at least some part of causality. It's still a correlation. So, whenever you see this disease, these alleles or these genes occur. But, they are what I would call good correlations, so there are reasonable assumptions, and there is even causality — so chorea-Huntington's a clear causality — you have these genetic repeats, and whenever you can count it, if there's more than I think 30 or so, you have chorea-Huntington, and you will develop chorea-Huntington, there's no question. Then, if you have a clue of what the genetic relationship is, you can look at the genes and find out what they are doing. What are they doing? Normally genes, the genes that we are interested in, they are the code for a protein — they construct a protein. And the next step is, what is the protein doing, in the body? And you can do all this experimentally — you can do this in animals: put the genes into a mouse, and then you let the mouse protein develop, and then you look where in the mouse the protein is, and so on. And so we have again a model, that is the famous mouse-model, for human beings.

Sorry to interrupt: so you will have a sort of identification phase which will be in a lab, just at the gene level, you just extract this gene and you work on it; and then you will have a sort of trial understanding phase of what is the matter, and then you will use the mouse to be able to…

Yes. So then you identify the protein and you look for where the protein is. And then you look for what the protein does. And if you are lucky, or unlucky, then the protein is part, or the receptor is part of the machinery normally on the surface of the cell that transmits signals from outside of the cell to inside of the cell. And if this is a growing signal, then the cell grows and divides, grows and divides, grows and divides. If you cannot stop this anymore, you call it cancer. This, of course, is not a normal state. So if you find that the protein is giving a permanent growing signal, you have to assume that this is a stupid situation for the body, because it has to grow and then to stop. So obviously the stopping signal is missing, or the stopping mechanism is missing. And that is the step where you assume that this is not the normal protein; it is a disease protein.

When you say assume, is it really an assumption?

At the moment it is still an assumption, because you do not know what the stop signal looks like. It could come from outside, it could come from inside, it could come from the protein itself. So you assume that it is a disease protein, so that the protein specifically belongs to the disease. And then you have one very attractive explanation: if this protein has a slight deviation from the normal protein, either in stereo or in electronic properties, which has been caused by a mutation in the gene, then you would have a logical chain to explain why this gene mutation did something wrong in the cell.

So, basically it's a matter of comparing a sort of static situation and what happens when you have a mutation; you let things propagate a little bit and then you compare.

Exactly. So then you look into the mice and say what is a normal protein in a healthy mouse doing? Is it at the same place? Yes. If it is at the same place, then you copy every two proteins, which is absolutely easy today — the computer does it, so we order the robot that does it, so you see where it is and then you see that there's something wrong. And what you normally find is an SNP — a single nucleotide polymorphism, where one of the bases, or two of the bases in the DNA is exchanged — by chance or whatever.

And so this is a way, what you are explaining here is somehow a way to understand better a disease, by going back to problems in the genes?

Tracing the causalities of the disease, the molecular causalities of the disease.

And the model will be here as a description of this causality — am I correct?

Yes. And then we come to the next step, where you see where the lock and the key is. So now you have identified the protein, you have identified the normal protein, you have identified the disease protein. And maybe the normal protein looks like that and the disease protein looks like that — it has some different shape. And then you say, ok: we probably could generate a sub-signal from outside, if we block the activity of the diseased protein. What we have to do is to consider this disease protein as the lock, and to construct a key that will interfere with the lock and stop it. Just prevent it from doing anything. So it would be compared to you putting in a key and then breaking it off, and then it's gone — nobody else can use the lock anymore.

Right, very clear.

Very clear, very simple, very mechanical. And that's the theory. So, the next step is technical, it's not science.

So, the model is basically leading to the construction of the key.

Yes, exactly. What you then do technically is, you take out the protein, say a milligram, you take out the genes, you put it in bacteria, you ask the bacteria to synthesise several milligrams or grams of the protein, you take out the protein, you crystallise it, you put it into an x-ray, you get a diffraction pattern you can calculate back from, and so on, you can formulate the relative positions of the atoms and then you get a 3D picture of the protein. Then you take your computer again and the rest is singing and dancing molecules, you only have small molecules, and then you ask the computer to look at the set of small molecules that would fit perfectly into this protein. And then the computer is doing something.

It looks so simple.

It is simple. And then the computer comes up with all kinds of suggestions, molecular suggestions, and then you can tell the computer well these are extremely stupid suggestions, but they already have intellectual properties of other companies, so please look at things that can be our own, so we can sell it; and the computer goes back and so on, and then you end up with a selection of molecules. So that's the easy part.

Ok, just a question before going back: let's say the computer comes up with different possible keys, how do you choose among these keys?

That's the first good question. In principle, at this stage of the project, you would choose at least 10 different keys, to develop them further. And develop them further means the keys should behave — and this is the other idealistic model — should behave, idealistically, like a magic bullet: going into the body, finding the target, destroying the target, going out and doing no harm to the rest of the body. Unfortunately, we don't have it. This was the famous Ehrlich paradigm [Paul Ehrlich]. And so the first selection would be simple: the imminent question is, 'is the molecule water soluble?', because 90% of us is water. There have been hundreds and thousands of molecules on the market that did not work because they were not water soluble. You cannot imagine…if I'm going to retire I'm writing about this. (At the moment, the most famous molecule or treatment for kidney cancer has been designed exactly as I have described it, and it turned out to not be water soluble. Then the management asked them, hey guys, are you stupid? What do you call these things that we cannot use afterwards, that we cannot use because they're not water soluble? Could you ask the computer to do something that is water soluble? And they said no, it's too difficult, if you want to make it water soluble, then you give it to the pharmacist, because that's their job, they do it, they make the pills and the tablets, they should make it water soluble, we don't care — we do the science. So then they gave it to the pharmacist and they said, oh that's easy: we add an OH group here and an OH group there and then it's water soluble. Now we have a little intellectual puzzle: if you add additional groups to the key…

It modifies the key?

Yes, they changed the key. So it did not any more represent the design. And then it was water soluble and they went back to the management. And they said now you guys we throw you out, you're completely stupid, you changed the design. But they had given it, in the meantime, to the medical doctors, and the medical doctors did not know anything about molecules, they simply put the molecules in rats and mice and humans; and then the management wanted to throw out the whole research department, because of stupidity; but the medical doctors called and said thank you very much, this is excellent, we now can treat kidney cancer.

So they improved the key?

They improved the key! This is very often the standard case in the development of drugs, and we call it serendipity. *Serendip* is the old name for siren and it's the legend of three little princes who wanted to march to Mecca, but they never got there because they got interested in all the things that were left and right; and they became enormously rich and wealthy and intelligent, and they never reached the far goal; so that's a good sign. You have to set the far goal, but in reality you will find something else. That again was the joke for Viagra, that was thought for heart and brain but it ended up some levels below!

But it's also being open-minded in a way.

Yes. So then if this process, and this is a nice example, would work like that, then we would have a water soluble molecule, and then the next one is it should not be toxic, at least to mammals, also, it should not be toxic to humans. And then economy jumps in, the market: if it is too complicated to synthesise… So, the most important department in a pharmaceutical company is a bunch of highly brilliant chemists who can reduce the number of steps to synthesise a complex molecule by 90%; so they go from 100 steps to 10 steps. And that makes it an asset for the company. And then you go into clinics, and this may change everything, because you have never had an idea what this thing does in humans, except your first idea that you met with this cancer. And then after the clinics, it's more or less legal work, marketing, economy, and so on and so on.

Right, that's very clear. So, going back to modelling now, I have some sub-questions to understand a little bit better this process. What would you say is the role of mathematics in modelling?

You cannot touch molecules, at least not normally; in CERN they can, but nano-structures we cannot. So you have to find a representation, if you believe in the lock and the key, we have to find a representation. And the representation normally looks like the following figure.

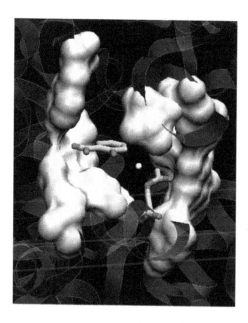

And the key, you tell me which one will be more meaningful.

So, this is aspirin, the activity of aspirin. That's the aspirin molecule…

So the aspirin is these little things and the big things…

This is the lock, and these are the keys. So, that is how we translate our imagination about the function of the molecule into something visible. And this is all mathematics. I mean, now we come into either cabaret or mathematics, I don't know. We know that atoms do not exist in a way; we know that localisation of the atoms in that way is difficult to imagine, also behaviour; so we have all kinds of reductionism, physical reductionism, to describe this kind of reductionism we need mathematics. So we assume, that's the physical reductionism, we assume that atoms are little balls; we assume that they have a weight; we assume that these balls are connected to a molecule atom by atom to a molecule by little strings. That's the most simple mechanical model; that's Newton mechanics, you need all the mathematics, that's not that difficult to solve it and make it a picture on a computer. Then you go to quantum mechanics, that's the rest of the story, so all this is based on quantum mechanics, where you include the electrons.

And then you have the problem that you have this nice Schrödinger equation but you cannot solve it; well you can solve it for hydrogen and so on, but that does not lead you very far. So then you have to invent all kinds of mathematics to have this approximation of the Schrödinger [equation], all these complex ideas that you have now, where you put the whole world into these operators, and so on, and so on. So, I would say…

So, you would say that the part of mathematics would be something like differential geometry, this would be really the key part?
Yes.

And besides this mathematical formalism and these mathematical tools, very powerful tools, what constitutes a model? Do you have other types of constituents — more qualitative for instance?
We still have in our minds, not the mathematics, that's the tool we have in our mind, the little balls, that are connected by strings or sticks, if I talk to somebody that is what is in my mind, that represents a molecule. And that is how we did it before the advent of the computer — I'm so old I played around with real mechanical molecules. They are [made] from wood…

…the red and the black.
Yes, the red and the black, they are from wood, and then some metal sticks, and so on, and so on. We know this is essentially wrong. But it's fascinating that it helps to understand what is happening, and to build up a hypothesis.

Can I just go a little bit away from this, because that's really interesting, and there is this sentence, and I think it was from a painter, I think from Matisse, which says that 'truth is not exactitude.' So it means somehow to understand…if you look somehow to be super accurate and precise, somehow you miss the big picture. So, would you say that applies here as well? Like somehow by having this mantle representation simplified, although we know they are wrong, we have a better big picture, or not really?

I think in this case — I love Matisse, but in this case — I would have another explanation. I think we are clear about the fact that we cannot find the truth, first. That's a stupid idea; we are hunting for the truth, but we will never be able to get there. So, truth and reality, *Wahrheit und Wirklichkeit*, are very different. No, it's kind of a question of coarse graining. Let me put it this way: if you want to handle an ice cube to cool down your drink, you would not need to think about all the water molecules in it and how they interact and what is the thermal dynamic state of a single volume; it simply works. You have the notion of an ice cube, you take it out of the fridge, it works. And so obviously we have not in every case to go down to the fundamental laws and issues to be operative in a certain way. That's the same way we do it with the molecules. We know that we have a lot of reductionistic assumptions about molecules, but we have some experience that if we treat them like little balls and sticks, then that's enough in terms of resolution. So we can make it operative, it works like that.

I perfectly understand, yes, very clear. Does language play a role?

Terrible. Language is the most terrible thing I ever… The day before yesterday, I talked a little bit with the speaker of the afternoon, Etihad Dzhafarov, a mathematician and theoretical physicist and Chair of Psychology — and he is interested in quantum-like behaviour of the human brain. In his recent paper, he has written a short paragraph on the abuse of language. And for a mathematician the abuse of language is if you do not use the * in assigning a certain symbol in a formula. This is already an abuse of language. I can understand it — by leaving out the little *, it completely changes the framing and view, and I can understand. And we have the same. So in talking about these things [points to a part of a diagram] or these things [points to another part of a diagram], it's extremely difficult to have a kind of a translation, to make yourself understood as if what you mean by that. And so we have to use all kinds of stupid simple things from daily life — lock and key, scissors, screws — all kinds of mechanical words where we accept that it is not true, but we have no other chance but to express it in that way. That is why we very often discuss, here in the Collegium, the importance of a proper language, because while we know that we express something wrong and that we have not the right word for it, we always fear that we are seduced into taking it for real.

Yes, there is a danger in using certain terms if it...

And so semantics is a big discussion here. Because you can imagine, between two disciplines semantics is terrible. There was a nice example yesterday night, when we discussed this [*with our fellow mathematician*], you immediately jumped to a different level. Because for her, she understood your language was perfectly clear. But if you discussed this with me or with a medical doctor, or with someone from *history*, it gets very difficult. And this is an old problem which is very nicely described for example in the biography of *Werner Heisenberg* in quantum mechanics, where he remembers that as a young student he was invited by Bohr, they were spending two weeks at a skiing resort, somewhere in the Alps, with skiing, drinking, sitting around, and mainly discussing the problem of language, because they had detected something new, quantum mechanics, with very clear mathematics, very clear formalism, very clear notation, but they had no words for it. So the mathematical formalism was clear, but now what does it mean? And they always said, we have to say 'particle,' this is the particle, but we know there is no particle! But how should we name it? You can name it Pauline, you can name it... I don't know. So this was a big problem that you always are put back to the classical view in describing a field that is opposite to the classical, completely... And so language is extremely important, and we are struggling a lot with language. So if *the mathematician* says something about body and if the psychiatrist says something about body, there is a complete different picture now. And theoretically it would be correct that in an inter-disciplinary setting like here, and in drug design which is totally inter-disciplinary from theoretical physics to the clinic, in discussing with your colleagues you would always have to specify what you mean if you say something. And I have met circles where this is done; in philosophy, they do it — they say 'this is a body' and then they say 'if I use the word body, then I mean this,' and so on.

This is quite interesting in your modelling, in the modelling process you describe — as you said, you have really different stages involving different groups of people; and so it should be like a flow of information, so language...

Yes.

Is notation important, or not that much? Because it's related — you mentioned this physicist and said the mathematical formalism was clear, the notations were clear, but there was a problem with language. So notation is sometimes not important, sometimes it's really important...does it matter?

Notation is really important also for the exchange of ideas. If you have clear notations, then you can transmit a message, and afterwards discuss semantics and the correspondence to the real world.

So, will it be like a pre-phase before language, you would say?

Yes.

OK, so now we go back to what is a model. Models are often said to represent a target system — so typically a selected part of the world. Does this characterisation describe what happens in your field? And if so, could you say how the model represents a target?

Yes. You already use the word which is very famous in molecular design — this is the 'target.' Coming back to this mutated cancer protein, that is the target. It is the same in the model, because it's the lock, and you need the key, and that is called the 'target' because it is the target of all that you're aiming for. So it's the far away target and you have to now, in an Aristotelian sense, construct a bow and an arrow and hit the target.

So that's a perfect description.

It's a perfect description. Somewhere I have a little bit graphically expressed what in philosophy is quantum mechanics, unfortunately it's in German,

I use it for a lecture. But Max Jammer's idea, he says — we have a physical theory here [diagram] — it has two parts: abstract formalism, and then correspondence to the real world. And the model is apart, it's on the side; it gives you an illustration of the physical theory, and then if you have part of the abstract formalism associated with the corresponding object of the real world, you come to a formalism that has some kind of reality relation. And that is interpreted; and then you have an interplay between…

Yes, back and forth…

Yes.

This diagram I think will be really nice, but maybe in English; that would be really good.

I can translate it.

That's perfect, thank you.

So, now we have mentioned a little bit of the context and theory, but what would you say is the relationship between the model and theory?

The easiest way that I would have is illustrating the theory.

So, you would have theory which is something bigger, and then based on the same theory or deriving from the same theory, you might have different models? A model being something aiming at different things?

The model would be my first…empirical hypothesis, to test the theory.

Coming back to this example of the protein, the lock: is it possible that two groups of researchers can have completely different models for the same target?

Yes.

It's possible — so completely different understanding…?

Yes.

And so they would have different keys, in a way?

They will have different keys. This is often the case because if the protein is mutated, the protein is not alone. Normally, these complex receptors are an association of several proteins; and they are surrounded by the membrane. And they are surrounded by a lot of other proteins that simply shut them off or put them on. And they even move them in the membrane. So there are all kinds of things… Normally, it's not enough to say we have this simple view. The next research study comes and says this does not explain what you see.

So the difference would be by the inclusion of other factors?

And, the position, for example, of very simple things — the positioning of the key. What we really did when I described the processes to you is, we took the lock out of the door…

And now you are including the door and even the wall! OK, I see.

And then it very often happens, that again you were right for the wrong reasons. That you put the key in the body, the key finds, at least it travels in the direction of the locks, and then it very often turns out that it interacts with a completely different lock. So *determinism* is really… And so a very typical example of that if you want to look this up probably is the insulin receptor. It started as being this size and now is like that; and now we have within one lock some hundred possibilities to insert keys and…

I think you already answered this question partly, but what would be the aim or use of the model? Learning, exploration, optimisation, exploitation…you might have different uses. What would you say in this case?

No, I think building up a model or making a model is a kind of illustration. So it is cutting out some reductionist part of the real world, and using it for testing your ideas. And then if you come up with a key from this model, then real life — *die Lebenswirklichkeit* — will show you what the key is worth. And that is why we still need clinical tests, because we are not able to predict the behaviour of the key in the real world.

Very clear. We mentioned this yesterday but I think this is quite an interesting topic, it's computer simulation: what is the relationship between simulations and the model? So you mentioned here that it plays a role when you try to identify at the beginning of the process...

Yes, that's the keys and they're inserted here, right? So this is a very difficult question. I have no perfect answer, but I have been thinking about this for some time. Somehow I would say that simulation is a real experiment. If I have a proper model, a proper representation of the protein — which are the blue lines here — in the computer, then it's 3D, it has a soft surface that can interact, so it feels like electronic interaction; it has a geometry which is hard enough that it is not destroyed, so it's stable; it behaves like a real physical entity. It's not identical, but then if it were, it would not be a model it would be reality; but it's on the way. And now I can do the same as I would do in a mouse, I can offer this protein, I can offer keys. And if the protein finds the keys attractive, then they will interact. And this has a scary reality. I mean it works. You can select from a hundred million molecular keys. You can select 10 that will also in reality interact with the protein. Probably not in exactly the same way, and probably not with exactly the same affinity or intensity, but they will. So usually the selections are not stupid. For us, then, this computer simulation is a kind of reality. We can trust that the molecules will come out in a certain way. And this process has even been automatised, so my successor there in the labs, in the chemistry department, he has built up an automatised lab, where you construct by computer simulation these images, you put in the keys, and calculate whether they will fit or not; if they can interact, you calculate interaction; energy, etc, everything. And then, if you are near enough and happy with the key, then you will hit the return button and then the computer in the lab starts to synthesise the key. And if you have assembled several milligrams and you again hit the return button, it will test this on cells. So within hours you have somehow transferred the problem from the virtual world into the real world, and then the tests of the cells, the results, these come back to you and are integrated again, or fed in the circle again; and then your simulation knows, oh no, this was not good, and they see that the amino group at this position did not have any activity, but nitro group at this position had this activity; so you make the system more intelligent or more knowledgeable.

I see the advantages of or the amazing thing about being quick and super powerful; but somehow it seems that being too quick rules out taking time to consider...

Yes absolutely; you're right, but I describe to you the actual thing. It is trying to beat the reality by lots of data. This was also a part of the discussion we had last night. So what we create here is big data.

Do you mean... I may say something that is completely wrong, but let's say we have two-dimensions, so time versus space and data. Before we were more in time, doing repeated experimentation, and now we are replacing this by being on the other axis and saying this should be equivalent in a way. Is this what you would say is correct?

Exactly. So I would have two other dimensions. I would say that on the one axis it's understanding and enrichment of knowledge — *Erkenntnisgewinn* — there's no English translation for that, it's a typical German. And the other line is collecting. Collecting was a typical scientific attitude, I think before philosophical interpretation of what was collected. So a huge collection of molecules, a huge collection of stones, or huge collection of plants, made you a hundred years ago a professor at the university. But to become a brilliant one, the next step was the ability to interpret the collection. So that is what we now have somehow transferred to the computer; and fortunately if you do it like that, you collect a huge amount of molecules and you establish cycles which have only one goal: that the key fits perfectly into the lock. And then you start an optimisation cycle that is easy. At the end, and this is also my criticism, you really do not know why it fits better.

And therefore you're not really learning for the next...

No. And with millions of molecules, you will not look up your lock files to find out which parts have been removed in the optimisation cycle. And that is dangerous because you have found a perfect key...

But you don't know why.

But you do not know…at least, you do not know what you have omitted under it. And therefore I think this is research which should not be called research, this is stupid mechanics.

So, in the long-term, what do you think could be the implications?

That is what I discuss very often with colleagues in chemistry and pharma, and they always invite me to do that, but the idea is to give a dinner speech, that they need a clown who tells them you are stupid, think first, and so on and on, but they do not change anything. And that is what they have understood — they are path-dependent. And that is a problem, because path dependency means you are on a certain track, a path, to find this key; and then in the first half of the path you are very successful, so you get positive feedback loops, because it works, it works, it works, it works. And then you cannot go off from the path; if you have been successful to a certain extent, a normal human being doesn't go back and say 'No, stop the whole project — first I want to see which molecules have been omitted by the computer and can we please look up the lock files?' And you would sit for weeks to inspect hundreds and thousands of data points; forget it. The company won't pay for that. So, you are somehow trapped by this positive feedback, going along the path and then ending up with a molecule which may at the end in the clinic turn out to be extremely poisonous to humans; and then everyone asks themselves 'we didn't do anything wrong, why does this happen?' But they did something wrong — they trusted too much in the simulation, they did not think enough about the molecule, they did not do enough critical thinking, have enough initiative, and so on. And this is a problem. So if you want to call it like that, this perfect world of molecular modelling is going to replace science with mechanics; or science with engineering if you want to make it a little bit more positive. It's engineering.

It's engineering. And so the big picture is…

The big picture is even much bigger than one could imagine. You could ask yourself, is the first assumption, that we should deduce that the existence of cancer is caused by the malfunctioning of a gene, correct? In cancer, this is more or less evident. But in diabetes, or in migraine, or in depression, even in infection, it's a question. So we have here inherited...if I'm digressing...

No it is really interesting, please do continue...

We have inherited the books and notes of Ludwik Fleck. Ludwik Fleck was a microbiologist in the '30s and '40s of the last century; he was German-Polish, which was terrible, of course — they caught him and put him in a concentration camp; he had written a very good book in 1935, *Entstehung und Entwicklung einer wissenschaftlichen Tatsache* ('How scientific tools are going to be established'). And this was taken up later by Kuhn, and he mentions Fleck also in the Foreword, but because it was 1935, the Nazis took the whole printing and destroyed it. So only three or four books — we have two of them — survived. But he was an epidemiologist, and he put forth the hypothesis that *maybe* in an infection the bacterium is responsible for the disease. But the question is, why does the single individual get infected — it has an immune system? So you need both — you need the consideration of the infectious particle, but you need also the health state of the individual, because normally it's not infected, so there is already something wrong when it gets susceptible. And so those were very much his ideas — he described the scenario of establishing an idea, a hypothesis, or a model of parts of science, as being like a river in a landscape: so it's always both, so the landscape determines the flow of the water, but the flow of the water also carves out parts of the landscape. So they do a mutual modelling of the whole world.

And it's also an interaction in both directions?

Yes. So only the action of both gives the whole picture. So, in this engineering approach of molecular interaction, the body's reaction is completely omitted. We have an abstraction of a human being, that you are a nice machine, and...

You have something fixed and then there is something moving in to it.

Yes, and then we can make all kinds of nice physical decorations; so we can move it a little bit, because we know about thermodynamics, then you shake it a bit, but this is not the movement that you have in the body; and you can capture it on film… But this is playing around — this does not change your basic assumption.

And those are the…enormous difficulties, given the fact that we are also talking about individuals, so even the genetics are not yet clear; of course, we are not so different in our genes, but still we are different because we look different. So we have all kinds of transcription factors and printing factors that make us different individuals. And how these play a role in what we would call a disease; in English there is a nice differentiation that we don't have in German and we don't have it in French I think: the Anglo-Saxon language differentiates between disease and illness. So the one is the somatic part — the somatic part would be the molecules; and feeling ill, this is somehow a kind of feeling, of mental state, and so on. Virginia Woolf's important book, *On Being Ill* — she has written a short essay about how there is a big difference in being ill or being diseased, and it's not so easy to explain to a doctor what illness is, and then I come back, sorry for that, to language, I can send you the title and it would be very important because she had thought about how to transmit her feelings to her doctor, because she said 'I'm not diseased, I'm ill.'

That's very interesting.

And the quotation I think is 'What about a language that has seen Shakespeare's poems and is not able to describe my headache in a proper word to my doctor. So I'm obliged to create new words like 'in-babble' from a part of poor sound and a part of pure meaning and lashing them together like a new coin — so coining a new word.' So she really…

Brilliant!

Absolutely, fantastic.

Yes, I would really be interested to have that, it's really nice. What is really interesting here is that there seem to be three levels: the reality; the computer modelling thing; but in the middle is the human modelling and what we can bring as… And by just going to the computer one we are overlooking this higher…

Yes. And you can analyse, probably I'm in a part of science that is very much struggling from marketing aspects, from the economy, of course. I can analyse all the links to the economy and it's very clear that these things come up. Because what you would avoid if your name is Novartis, what you would avoid is to make therapeutics for a single person, unless this person is able to pay. What you would avoid is to have too many philosophers in the development pathway, because normally now it's one decade for a new drug, and you would probably have two decades, or three decades; you have to avoid synthesising too many molecules because you are not allowed to produce waste — that's from the regulators, that's the law. So everything that comes out from the waters in Basel is clean water — nothing else. Waste is a real problem. So if you make too many molecules, then you have a problem — you have a law problem, a regulation problem, you have a financial problem, and you have an organisation problem, a logistics problem, it's terrible.

It's modelling under constraints, and so the constraints are...

Yes. And making virtual models, and interfacing them in a most clever way to reality is also an economical part of a company. So the more you can do *in silico*, the less money you have to spend and the less you have to fear the lawyers. And so it's an extremely complicated picture that you can build up — a multi-dimensional relationship. So it's pretty complicated. And that has political consequences. If you cannot do it in Switzerland, you go to India. And if you have enough money, you always find a way to do this; or in Mexico. Modelling in Mexico is completely different, because there are different ethics. Here, ethics influences what you can do.

And how creative you can be.

In Mexico, you can persuade some volunteers.

So do you think that in very developed countries we are going towards engineering, and in less developed countries, that's where science will be?

No. We are going towards engineering in very developed countries because of economic pressure. Nothing to do with development; because this summer I travelled a lot in Asia and visited all kinds of universities. So creativity is another business. Singapore has a huge building, huge, fantastic, called Create, because they have realised that none of the other clinics at the universities are creative at all, because they are all doing modern technology — what they do is engineering.

I think there is a question which is appropriate now: what has the impact of the development of new technology and tools been? For instance, in cosmology, the introduction of telescopes. So here, probably computer simulations?

Yes, for me that's a very, very clear development.

It seems to have completely changed the nature of research.

Yes. We have in my field a co-evolution since 40 years, 50 years ago, of molecular biology, biotechnology, computer techniques, automatisation, and mathematics. And these fields have transformed each other, because with the aid of mathematics and simulation, these things were possible. With the aid of mathematics and automatisation and robotics, you can do very many experiments that are provided from biotechnology and molecular biology. Molecular biology has the advantage that you can simulate a cell in reality, not simulation, but you can take out parts of the cell...

Just one cell in the...

Yes, just one cell or parts of the cell. And that makes the experiment extremely quick. In one night you can do two million tests on a hundred different cell systems. It's an automatic lapse. It has another economical dimension; scientists don't like it because every test is one dollar. So within 24 hours, you've spent the budget of the whole year, and they are very reluctant to do this. We'll come to that in a minute. So this has completely transformed and shaped biology, pharmaceutical medicinal chemistry, medicine, it's gorgeous. And you cannot turn back the wheel, that's very clear. But we lost a kind of proper classical deduction process, we lost a kind of induction process...that's very clear. We shifted things to engineering. And we hoped that by simply considering many data points, somehow we would see the emergence, that something pops up and suddenly becomes clear. And I doubt this, because behind all these developments there's an algorithm, and it's only the limits of the algorithm that you see, it's nothing beyond. And whether this is creativity I doubt. The funny thing then in this transformation process is that we have two or three points of human interaction that I find very funny, and this is not published, it's never published because I only have it from the sayings of my colleagues: let's take the example of the two million tests per night; nobody's going to do that because they need to spend all the money of their lab and they have to risk that they end up with nothing, or only 10 molecules or something like that. So this guy tells me, no we reduce, instead of two million we only take 20,000. And I asked, what is your criteria for the 20,000 out of all of those two million? And they tell me 'intuition.'

What is intuition? How do you get it?

They do not know, but the experienced medicinal chemists have a feeling that for this lock…

There is a threshold.

…the key should have a nitro group there and always good to add iodine there, and so on. And then they go to the computer and they have the two million molecules to start with and they extract 20,000 — they say 'show me all molecules with iodine and all molecules with a nitro group there,' and they extract a subset. And very often, they are right. And nobody knows what this is, this intuition. The head of Novartis, for instance, would love if I could give him an explanation of what they do. They don't know it themselves; they call it experience. They say 'we have seen so many molecules in our life that work and don't work, so probably in our subconscious brain there is some imprinting pattern, that gave us an idea of how they would work. And this is very frequent behaviour in… And for me this is kind of ironic, because you have established all these rational algorithms, and then at a certain point, a human being interferes, with very questionable methodology, just for the sake of economy. It's crazy.

That's quite interesting.

It's absolutely crazy. You as a mathematician would say that's probably the most stupidest thing…

But you know people in financial markets or people practising sometimes come up with what is called the rule of thumb, and it makes, there is something there…

There is something in it. And these old guys, very often they're right. And they have to…

Just a feeling, experience.

And then the other way, where I see the whole system breaking down a little bit from the borders, or from the boundaries, is that we have now more and more complaints from basic scientists, so let's say yeast research, yeast genetics, that they are all transformed by modern technology, either by robotics, quick experiment, and so on and so on, that one of my colleagues, one of the superstars of yeast genetics, says he has a little farm in France somewhere, and he leaves the lab and he goes there for six weeks, and says 'too many experiments, I have to think about them; I do not understand it.' And he feels that he's transformed by the medical demands — so whatever he does, whatever he publishes in basic yeast genetics is automatically sucked up and used by medical application, because the money that is in medical application is the most. And he says, 'all my research, all my thinking, is going from my surroundings and going to be transformed in this medical direction.' He says 'I'm not interested in medical data, I'm interested in yeast genetics.' And so this is the danger — that we face very strong pressure from the utilitarian view of basic science. And he feels it very strongly. So he says 'whatever I do, the yeast gene people interpret it in terms of being applicable to cancer research.' And he said most of that is stupid; and 'if it does not work, they say I was wrong, but I was not wrong, I was doing yeast genetics, I was not doing a model for cancer.'

Do you think that some very big discoveries of the past would not be possible today?

I fear that, yes.

Because of the way it is…

Yes. People do not lean back any more and think it through till the end. That is, by the way, one of the reasons why my president wants me to do this scripted thinking initiative, because he says we make too many experiments. Wait a moment, look at it. So that then is also my criticism of too much computer simulation, if we can automatise all the simulation business and interface it with reality, as I describe to you the automatisation layer, then we will avoid leaning back, because we get permanently positive feedback. And only the negative feedback wants you to lean back and… So this Gottfried Schatz, the former President of the Swiss Scientific Council that I'm going to take over now, has said the goal of science, at least basic science, is not the truth, it is not the knowledge, it is that which we do not know; it's the non-knowledge, to shed light on the non-knowledge. And conducting experiments is one methodology, but it's not the goal of a scientist to do an experiment that gives the right result; 'if we use them,' he said, 'we should always see some experiments that give us the results that we never expected; then you will learn something.' So I can understand that very many of my theoretical physicist colleagues were a little bit upset about the Nobel prize of the Higgs Boson, because they say, 'oh it fits into the standard model, how ugly.' So we are still in the state that we can only ever explain 5%… The nice thing would have been if they'd destroyed the standard model and then we could have started from the beginning. And I can understand that.

So, what is a good model?

Well, I feared that.

Everybody does!

In principle it's easy. The question has not been solved for French companies like Dior and Lagerfeld, but I think in my field a good model is very clearly defined. It's the physical (analogue) modeler computer model that enables you to design keys that have relevance for clinical purposes. And then you don't ask how simple the model is, or primitive, and so on; if the model is good enough that your keys make it in the clinic, that's the important thing.

OK. So, now we move away from modelling, and we go towards the notion of risk and uncertainty. I make a distinction here, but maybe it does not apply to your field. I start with uncertainty. So, how would you define uncertainty, and how does the model help us understand uncertainty?

My first question is, is uncertainty uncertainty, as in *Unsicherheit*? I ask for a special reason: again in quantum mechanics there is the Heisenberg uncertainty principle, and there is a big discussion in philosophy, and very many people think that uncertainty is the wrong notion, it's the wrong translation. What Heisenberg meant was *unscharf*, which means 'not precise enough.'

That's what an economist would have — exactly this distinction.

Uncertainty is wrong, it's not uncertain; it is absolutely certain, if you calculate it, the particle is there; it may be not a particle but a wave, but it doesn't matter, it is there, it is not uncertain. It is not precise enough because you have a technological problem for measuring; so if you want to make high precision measurements, then you get not only one sharp point, you get 15 points that are scattered around the real point. And then you have to do probabilistics or statistics. And that has been translated wrongly into English into uncertainty, but this is not uncertainty, it is lack of precision. They are two different things. You are not uncertain about them; your statistics tell you with 99.9999% certainty that the electron is there.

So, I guess this will be more about whether there is one or not?

Yes. And by uncertainty what you mean is that you are uncertain about framing the problem, explaining…

Different fields have completely different meanings. I guess the way I rank them, if I could, would be that risk will be better defined, and uncertainty will be opening room for really things that you do not expect. If that makes sense.

Then I would agree, because very simple-mindedly I would say that risk is something that I'm going to calculate, either with mathematics or with my gut feeling. I might say, 'oh that's too risky.' Uncertainty would cause me to stop anyway for a moment. So it's a completely different category.

Yes, that's much above…

Much above, yes. I would not take the risk to be uncertain about something.

Somehow, my feeling from our discussion on computers and everything would be that, risk would be something that computers could do and estimate and calculate, but uncertainty is something that the computer will go completely without looking at.

Yes. Risky behaviour, I would say, if we come back to the humans, is understandable. Behaviour based on being uncertain is kind of stupidity. If you are not certain that the ice on the lake is thick enough that you can stand on it, then it's not a question of trying it, you are simply uncertain; it's a different class. If you look at it and you are a person who takes risks, then you say, 'ok, maybe a nice part of cold water that I easily stand on for some minutes may break in; it is worth trying it.'

So, would you say that the model in your field will help to understand both, or will take into account both, or not really?

I think these are two different classes. If I talk about these molecular models, I would talk about uncertainty, not about risk. I mean, we do the models because we are not sure; we have uncertainty as to whether we even frame the problem correctly. And that's nothing to do with risk. It's not risky to do the simulation, it's not risky to do the experiments with mice; but we do it because we're not sure whether this protein is really responsible for something. So it's not risky it's…

It's really uncertain?

It's uncertain.

Just a separate question: does it happen that no key can be found?

Yes.

Because of the progress of science at this level, or because no possible key will ever be found?

No. I think it's the progress of science. I'll give you a very simple example. Again, back to the model: the real locks and keys are flexible; so you can imagine it's not that easy if you have a flexible key and a flexible lock to open a flexible lock with a flexible key. There is a very stupid but simple technology to avoid this for modelling: you freeze it. Then you have stable locks and stable keys. But that is not the real situation.

No.

So, what we have not understood is the turbulent entropic behaviour of the lock and the key. And therefore, in a frozen world, you can design a perfect lock. If it then comes to the real world, sometimes the lock and the key do things that make you crazy. So the interaction of the key can generate a complete conformational turn of the lock; so it will change the shape. And then, you like every explanation of what's going to happen; and then you will never find the actual state of science, a proper key, because we cannot interpret what the thermodynamics of the receptors are. There are people who are working on it. Thermodynamic or entropic control of protein movements is a relevant topic in physics.

Very clear. Now the last question, which is probably the most personal question. For you, what is the experience or the result in your field which has had the most significant impact and why? It can be your own research or it can be things happening in your field.

For me, it is probably protein crystallography. So we understood from former colleagues, the generation before me, that crystals are not necessarily dead matter, but that we can have functionally active proteins in a crystalline state, which means geometric repeats, that is then accessible by spectroscopy, so that we can do structure simulation. And that was the first time that we saw that all proteins have common features — they have modular structures that come up over and over again, like these helixes, and so on and so on. And this very deeply changed the understanding of the locks…

Because then the structure of the protein was better understood and it was easier to…

For the first time we had the structure, and we knew what our target would look like. And then the field really exploded. And I remember this so well, because it struck me, as a very young PhD student, that the Californian biochemist Dickerson wrote a biochemistry book which is as such not exciting, but he teamed up with a guy named Irving Geis; and Irving Geis was one of these wild hippy sprayers who did excellent graphics, and he asked him to illustrate this biochemistry book. And it was full of helixes and turns and whole pages of imaginations of what these proteins would look like. And I stumbled over this book and I said 'I want to do this,' it was just so fascinating. His autobiography is very personal, so it really struck me that now in this world we could build something, understand something.

Chapter 11
A Conversation with Leonard Smith

Can you describe your field briefly?

I would say we work at the interface between models and reality, trying to get observations of the real world into the model-land, and then interpreting simulations in model-land so as to say something about the real world, usually something about the future. My work draws from many different traditional disciplines and often focuses on nonlinearity, on how the mathematics of nonlinear systems differs from that of linear systems. If there is a common thread, I guess it is in the attempt to see to what extent our best theories, in combination with carefully chosen observations, tell us about reality.

In physics?

Not just physics, although physics does have the long running advantage of repeatedly redefining itself to keep the focus on things where today's laws of physics are informative! In many ways, physics is a good place to test a new idea first. That said, one of my students is in the Bank of England's "model risk" team, one has been awarded the US Navy's highest civilian award for building a model to predict pirates, one focuses on analysing medical time series, I think one who went into the city has effectively retired! Others are professional authors, working in an applied sector or doing physics or applied mathematics in universities all over the world.

Present a paradigmatic example of a model in your field, describing it in terms that are accessible to non-experts.

The prototypical example of a big model would be a numerical weather prediction model. Weather models fill up all the available space in the computer that allows making the forecast on schedule. But let's start with a small model. For example, the following rule: the new value of x is equal to the old value of x, times one minus the old value of x, times a constant λ, where λ is a value between zero and four. (Four is by far the most popular value!). This mathematical rule is called the Logistic Map, it was proposed during the Second World War as a pseudo-random number generator.

So, the equations behind this rule constitute the model?

I think of a model as the actual thing that generates the output, the simulation. If you do compute things by hand, then the rules you follow (and the mistakes you make!) define the model. Today, however, the model usually consists of a computer implementation of some ideas, mathematical ideas chopped up to run on a digital computer. And the particular computer itself, I'm afraid. The best weather models tend to fill the biggest computers of the day, taking today's laws of physics plus whatever else you need to make it run, then looking at what is happening at the moment in as many different ways as you can (collect observations), translating those obs into model-land snapshots of the current weather, and running that initial condition forward — minutes, hours, days, months, or years — and finally interpreting those simulations to form a forecast of the future weather in the real world.

So, would you say that you have things, like the laws of physics, which are not a model, but true relationships between different things in physics, and what you're looking at in a weather model is the link between this and weather conditions, in order to make forecasts?

Yes, an operational weather model is a link as you say, but I hesitate to use the word "true" in science. Today's laws of physics are in a sense nothing more than an analytic conjecture. Mathematical analysis of these equations can tell us fundamental mathematical constraints we were unaware of. They tend to be phrased as analytic models, which are mathematically rather different from computer models which merely shift bits about.

Can you explain what you mean by analytic model?

An analytic model is just a set of mathematical equations. Like the Logistic Map. The "unreasonable effectiveness of mathematics in the natural sciences" is a common theme in popular science, but from a mathematical point of view, our models are far from exact (outside number theory, the integers). Wigner, who wrote a paper with that title in 1960, hedged his bets more carefully than most modern writers. Many people today view the laws of physics as "Truth," but physics doesn't do truth, you know — religion does truth. Physics does "super-cool approximation." Once maths goes beyond the integers (and mathematics underpinning weather forecasting goes well beyond number theory), the laws of physics are just "today's laws of physics." For centuries we believed that Newton's Laws were true. They are not, and thinking as if they were stymied the progress of science significantly. I believe we run a similar risk with computer simulation.

So, would you say that an analytic model is an equation linking together different quantities?

Yes. Quantities. Or concepts. Phenomena we know about. Abstractions representing these phenomena go into the equation(s) we write down. But the concepts we have differ from whatever it is that is "out there," the unknown thing philosophers call "the data-generating mechanism." Analytic models aim to capture what we know with sets of equations; even if they appear to succeed, we know we don't know of everything.

So, what you are saying is that we have an approximation because it is simply based on what we know?

I hesitate to say it is an approximation, since "approximation" makes it sound that there is a "true" mathematical model out there which we are approximating! I often doubt that mathematics can capture the thing that's out there. But yes, we are limited to what we know. I find it helpful to think of simulation, if not all of science, as apophatic science: whatever is really out there is simply not something we can conceive of.

Fred Hoyle, a former Astronomer Royal, wrote a fictional account of a scientist who tries to comprehend an arguably ultimate true physics; the story does not end well for the scientist.

Will representation be a better term?

Yes. I like representation. Or even "model" in the snap-together plastic model aeroplane sense.

So, an analytic model involves some equations which specify relationships between different concepts we know. And you're saying that the models that you're interested in, you mentioned the weather model as a typical example, will be more on computer experiments, simulations.

That's right. Of course, being "analytic" isn't enough; we have mathematical models in finance, and ecology: ideally you need a really, really good analytical model, like Newton's Laws of gravity, a model so good, you can make the mistake of thinking it is "True." And of course, it is not.

In the physical sciences, we actually have models that are that good — models that we believe describe exactly the way things really are. But of course, the way we think things really are is contaminated by looking at them through our models! Regardless, our model-things look so close to the real thing that it is easier to make this mistake than in some other fields, and inadvertently believe that our mathematical models are more "unreasonably effective" than they are. In a mature field, this is deadly. In a developing field, well I am not sure if it helps or hurts you. Certainly, you want to think your equations will work better than the old ones you are competing against. How do you decide what's better if you do not believe in a "true"?

Is it a verification of the laws?

We never really manage to verify; but we repeatedly evaluate. We do succeed in showing that our equations are non-trivially inconsistent with the observations. But nonlinear equations are hard to work with analytically. This is true even for Newton's Laws: we can work out the solution for two bodies, we cannot solve the equations for three bodies analytically. We can solve the Earth and the Sun, or the Earth and the Moon, but not the

Sun and the Earth and the Moon. And when we put the equations on a computer and simulate, well computers don't do real numbers or continuous fields, we have to make approximations. It gets harder even to show it is the equations that are wrong!

But what about the model? Just going back to the concept of the model, the model will be the relationship between the Sun and the Earth, and the Sun, the Earth, and the Moon?

Consider physics as a solar system with three planets. We write down the equations; we believe those are *exactly* the equations, but we can't *solve* those equations. So there is an issue of tractability. And *then* we make a computer model. One could call the equations a theory and the computer realisation a model, but they each feel like models to me.

So, would you say the models you are working with are built themselves on models? So it's like a second level?

I'm happy to say that Newton's Laws provide a complete analytic mathematical theory. But it is not possible to solve these equations analytically. So, we then build a computer model, a simulation model.

Maybe it is a good time to talk about the difference between model and theory, because it seems quite important for us to understand. Does it mean that you have a theory: for you the difference mainly between model and theory, is that the theory comprises very general laws, that are not necessarily tractable, but then you build a model, based on this theory, and the model adds the tractability?

Correct.

And it's more specific. For the same theory, you will have different models built depending on what you want to understand?

Perhaps. You compose your question within the theory, and you may or may not be able to solve the mathematics. Alternatively, you can build a model to approximate the theory, answer your question in model land and then wonder what the model land answer tells you about the actual answer. Ideally, the theory would cover everything, whether you wanted to understand everything or not. For reasons of tractability, and the limitations of today's computers, different models might focus on different phenomena depending on your target. But even then, the theory isn't Truth, right? Even then the theory is also only approximating, representing reality. So some people might call the theory "a model of reality" and the computer model "a model of the theory," and some people confuse all three, thinking they're all the same thing. When confusion sets in, it is often helpful to learn if someone believes today's Laws of Physics govern reality, or mere describe it.

So, that's why you would have a model for weather, but not a theory for weather?

I'd rather say that many different theories go into the model for weather.

Different theories?

Different bits of theory from different bits of science. Suppose we made up a planet like the Earth but completely covered in water, and assumed the Sun was just as it is now. To model such an aqua-planet we would need to include the effects of a gravitational theory (to get the orbit, tides, and seasons accurately), a theory of fluid dynamics (to get the ocean currents and waves), a theory of gas dynamics (to get the atmosphere) a theory of radiation... You would have to decide what to leave in, what to leave out. How much code you can write?, how big your computer is? can you afford to include ocean surface waves pushing against the wind?...

What is clear is that things get too complicated to work them out in our heads or by hand on the blackboard. We can work out some analytic consistency checks in our heads (conservation of total water, conservation of momentum...) but we cannot work out the weather analytically even on our simple aqua-planet. The equations we know we would need make things much too complicated. So we go to a digital computer.

And the computer model, what is it exactly? Is it a code?

That is a great question. The short answer is no.

Let's suppose we want to solve a nonlinear ordinary differential equation on a digital computer, Newton's Laws are a good example. Suppose we wanted to know when and where the very last total solar eclipse on Earth was going to be. The computer requires we discretise time, to take small but finite time steps. Simulating a continuous field like the ocean, or the atmosphere, gives us a partial differential equation and then we must discretise space as well. And a digital computer always chops our real-valued variables into finite precision all by itself; you are not forced to think about this, but it will affect the accuracy of the outcome. In any event, we are no longer solving the original theory, we're looking at simulations from an approximation of that theory.

So you do all this and come up with a few thousand lines of FORTRAN code. Is that the model? Well, no. If you run that code on a different computer, or even the same computer with different compiler settings, you will get a different simulation for the same inputs. I'd argue the model has to include everything required to get exactly the same simulation. But even relaxing that demand, the model is much more than the code.

And so would you say that there is this step between theory and model? It seems that theory comes before a model and there is this step about making all these assumptions — about which time step you use, which step in terms of space discretisation you use — so you have an additional phase which will be the assumptions phase. Will this be the first step in the model?

Traditionally, yes. Although I now see it as more like a series of slippery slopes than a step. There is an entire field of numerical analysis which tells you what time step you need to be so-and-so accurate for so-and-so far into the future. In complex nonlinear systems, it is difficult to know how long your computer code can shadow solutions of the equations you were aiming for, much less observations of the physical system. Once we move to observations of the real world, we can happily look and see how long the computer simulations can shadow the observations. This is not forecasting, rather it is asking, _after_ you see the outcome, whether or not your computer model can reproduce it: can you shadow the observations once you know what they are? Knowing you cannot get it right even when you know the target helps to determine how to interpret simulations as forecasts, and when not to. So, there is the question of fidelity and there is the question of shadowing.

Let's go back to the weather model, this computer model. Can you explain to us why a model is needed and what a model is? So, mainly what I have in mind here is not only the ultimate objective but also the various stages in the modelling process. So [firstly], what is the role of mathematics in modelling?

In modelling physical systems in general, and weather modelling in particular, the fundamental conceptual task is to integrate the equations of motion, and that is to solve a maths problem. Since the 1950s, when people first attempted this solution on a computer, the limitations have been technical: we can write down equations that are too complicated to solve with today's computers, so mathematics also plays a role in deciding which simplified set of equations to try and solve. Third, mathematics helps you decide how to evaluate if your solution is any good, both *a priori* as a mathematical solution of the equations you were trying to solve (an evaluation in model-land) and also whether or not it proved to be a useful solution in terms of the weather, an evaluation in the real world — of course, this second task calls in a mix of pure mathematics and very applied statistics.

For 40 years, weather forecasting felt heavily constrained by the available computer technology. We knew more physics than we could include; and the available computer power wasn't able to do what the numerical people said one needed to do. While that is still the case, for the last 30 years we have judged our models good enough to invest precious computer time in quantifying the uncertainty our model sees in the future. Mathematics plays a major role in how we sample that uncertainty.

Apart from this mathematical formalism of the model and the computer simulation aspects, what are the other constituents of the model?

Ideas and concepts. The physics. Of all the phenomena you know of, how much computational resource to devote to each. How detailed a radiation code against chemistry; how to balance aerosols against the wind? What do you think is required to get high fidelity model simulations?

So, this including or not of a certain aspect, do you do it by trial and see if you are sensitive to inclusion of this particular aspect, do you do a "robustness" analysis?

You would like to, and when modelling "simpler" things, like laboratory fluid dynamics, we can do exactly that, even changing the situation to see if the phenomena we think are important actually guide what actually happens in reality. Moving to weather, however, well first, it is harder to do controlled experiments and second, our models are still relatively coarse compared to things we know to be important. If a model cannot see anything smaller than 10km, how do you account for the drag ocean waves have on the surface wind, or giant tall clouds which are much smaller than 5 km? Improving the resolution of the model eventually overpowers the computer, and even before that it slows things down so that the simulations are not ready for the BBC at 4:30.

So, there are these practical constraints when you're doing operational modelling. In theory, there are other ways, you can look at the model just to look at the model in research, but that's extremely expensive. Ideally, we often do what you suggest, but in practice, we try one thing or two things and if it looks ok, we go ahead.

Time constraint is an important constituent in modelling in this case; what about data? Are they important?

Data are critical. Both the observations that tell us the current conditions in the atmosphere (the "initial conditions") and the studies that determine parameter values and indeed the mathematical form used in the model for various phenomena are critical. Traditionally, weather modellers insisted on "high quality" professionally standardised measurements. Today, there is an interesting question regarding whether or not the vast quantity of "low quality" cloud-sourced data can best be used, say, to evaluate our forecasts if not to initialise them, and why not to initialise them as well? I expect that as we think more about information content and less about pin-point accuracy, entirely new sources of data will improve our ability to forecast and to understand.

I would say there are three critical areas: one is how good the mathematics itself is, whether or not the mathematics reflects to whatever is out there; one is whether or not the models we develop from the mathematics are actually tractable — whether or not we can solve them at a reasonable resolution; and the third one is this question of what the current state of the system is — issues of noise, or observational quality.

I think here, if I may, I will introduce a question which is not in the original list. It's quite interesting because from my background, which is more financial risk or insurance risk, usually your starting point is known, and then what is unknown is the dynamics; so it's more stochastic. Whereas what you seem to say is in a way the evolution tends to be something that you have a better understanding of, but the problem could be understanding where you start from. Can you say a bit more about this?

Sure. I think there are two issues here, one hinges on what you believe, the other on what can be shown to work well. Ernst Mach argued 150 years ago that the dispute between determinism and randomness in physics was a distraction, since it could never be decided by experiment. In weather modelling, the best equations we have for the large scale motions of fluids are deterministic, as are all the serious contenders. While there are many open questions (millennium questions!) on the behaviour of these equations at the smallest length scales, that is not relevant to weather forecasting as we just cannot reach anything like those length scales. So a completely different question, a pragmatic question, is whether you treat those things you cannot simulate as constants (the mean of a process) or as stochastic realisations somehow similar to the things you are leaving out (so-called stochastic parameterisations).

One can believe in a completely deterministic world and still defend the use of stochastic parameterisations in a computational forecast model.

I'd also like to disagree slightly with the question's statement, before coming back to agree with you. If you look at the quarterly report of the Bank of England, you'll see that when they look at the history, when they're starting their models, they actually plot their uncertainty in the past. And next quarter they go back and note that the uncertainty with today's "past" will have decreased, and this may impact their thinking next month. So I think initial condition uncertainty does play some role(s) in finance. But on a more fundamental level, you're right — there's sort of a division in physical sciences' modelling, between people who believe "the truth is out there," the X-files guys, and that the uncertainty is only due to our ignorance. Or to paraphrase a Frenchman[1]: if we really knew the exact state of the weather right now, and we had a super-fast computer with unlimited speed and precision, then we could make forecasts that would tell us almost everything we wanted to know, and those would be point forecasts. Even if we do not know everything precisely, the X-files guys often use the work of the same Frenchman to put probability distributions on everything we are uncertain of, and then make probability forecasts. I hold more of a "The Matrix" position: "there is no spoon"; it is a harmful distraction to act as if the truth is out there, in applied forecasting it is much more effective to embrace model inadequacy as a fundamental aspect of model-based prediction. Always.

So, this sort of depends on what you believe. But one of the roles of mathematics is to help you figure out whether or not you believe something that is silly, or wrong.

So, just to be clear: by point forecast we mean the temperature in London tomorrow at 12 o'clock will be x degrees, whereas with a probabilistic forecast you have a distribution of possible temperatures with different likelihoods for tomorrow?

Yes. The temperature measured at London Heathrow Airport will be 12.1415...degrees and it will be raining. Or a probability distribution ranging from below zero to above 30, but with almost all of the probability mass between 11 and 13 degrees. (And it will be raining with probability great there 99.99%.)

---

[1] Editors Note: That Frenchman was Laplace, story of demons in *A Short Introduction to Chaos*, by Leonard Smith, OUP.

Are you an "X-files" person or Bayesian person?

Neither. In practice, I am more of a "there is no spoon" person, from the film The Matrix. Our models are both fundamentally useful and fundamentally flawed, flawed in their mathematical structure. So there is always a chance that the real world might do something you cannot see coming from a model. Indeed, there are always things coming in from outer space that even in theory we cannot see coming even if our model could handle them. You might say that I am a pragmatic Bayesian, I would like to frame the problem using probability theory, I know I cannot solve that problem, but it guides my thinking. There used to be some Bayesians called "Doogians," I'd like to be one of them, probably.

Does language play a role in the modelling?

It does, at every level. We say our models are deterministic, but if you change the computer, or the compiler, then the same code will give you a different number (if it runs at all). This is not deterministic in the sense that a set of equations are deterministic. Given a set of equations, you just look on the right-hand side of each equation: if there is no random term lurking there, then the equations are deterministic. Our models are not deterministic in that way.

Of course, small differences in the initial conditions might grow rather quickly, even when the model is deterministic; thus "chaos."

Also, when we try to take into account this fact that we *don't* know the initial conditions, even if we believe the best model is deterministic to the extent we can do that, we actually sample different uncertainties in the initial conditions. And since 1992, the US and Europe, and soon after the UK, and now almost everybody samples this growth of uncertainty within their model via an ensemble of simulations…these are ensemble forecasts where we really try to explicitly take into account the uncertainty in the initial conditions, underneath this deterministic model, but the forecast — we get a probabilistic forecast — is really a question of trying to trace what we mean by uncertainty and where it comes from. So this idea that we don't know exactly where to start, so we get slightly different answers.

We will come back to uncertainty later, but that's very important...the fact that if you have to communicate a weather forecast, are there any specific problems in communicating a probability forecast?

Specific challenges, yes. Fundamental problems: no. At least not in Britain, and I expect not in Europe or America. There is a sign on the bus stop just outside this building. There are two photos: a serious athlete on a silly bike, and a scratch card. Over the first are the words "No Chance," over the second are the words "1 in 4 Chance."

I have seen many adds like this over the last 20 years. Adds on the highway which say nothing more than "8:1 there are no workmen in the next designated road works." Ads which assume a basic understanding of probability and odds are fairly common in British society. Now that the Met Office forecasts are not distributed by the BBC, it will be interesting to see if they shift to communicating probabilities. The New Zealand Met Office have been doing this for years, and I expect it has saved lives, is saving lives. The Dutch have put probability forecasts on television; in the US, people have given the probability of precipitation ever since the early days of forecasting, long before computer forecasts were competitive with human forecasters. So, I think the key is that we need to be clear about what we're forecasting; and very often people find ways of using that.

Another key is to be transparent and consistent in the long run. "Rain" in Germany is defined differently than "rain" in England; you can have the same storm called "a hurricane" by the Australian Bureau of Meteorology and "not a hurricane" by the Japanese Meteorological Association. I expect clarity and consistency are the big challenges for those communicating model-based uncertainty regarding our future.

For weather, we get similar forecasts over and over and over; people learn how to use that information, even if they don't use it the way a mathematician might say they should, some make better decisions consistently doing so. Of course, for climate, the situation is rather different.

How important is notation?

I do not find differences in notation as challenging as differences in what people interpret the symbols to represent. Differences in notation sometimes indicate difference in representation, but more often I find using the same notation hides important differences. I talk to people in many different fields, and I aim to be a jargon chameleon. As long as the concepts and the notation are clear, as long as it is agreed, then I don't find the notation particularly important. What's critical is to realise that x in a model is different to ~{x} in the real world. So we may call it temperature at London Heathrow in the model, but that's a different thing than temperature as read on a thermometer. The only time notation is a problem is when it leads to ambiguity.

Models are often said to represent a target system. Does this characterisation describe what happens in your field?

Yes, I think so. Definitely for weather. Of course, there is an issue in decision-making when (at least all but one of) the forecasts are counterfactual. What does it mean to forecast a policy path not taken? Or for that matter: should a probability forecast get any credit at all for the probability it did not place consistent with the one actual outcome? How do you evaluate a policy target to achieve a 50% chance a "bad thing" will happen? In any event, making a forecast for 2300 which depends on what we do today is not really forecasting the future, it is trying to aid the decision about what we do today. Confusing our imperfect models with reality in this context can be earth-shattering.

How do you understand the model–world interface?

In general, we understand the interface "badly."

I understand it probabilistically; I think of it as mapping observations (like a thermometer reading at London Heathrow) into model-land x. You have to build a map, a mathematical map, a guide, that is called data assimilation. Then you make your simulations and, if you think a particular simulation or simulations might be informative about the real world, then you have to get back out again — you have to come back out from the model-land to the real world. Arguably, that is never simply reporting what the model says; especially, even when things are going well, since a point in model-land corresponds to a distribution of different states of the real world, not just one. And often, things do not go well.

Half a century after computers became commonplace in academics' offices, naïve interpretations abound. Even to the point of saying things like "the problem with the observations is that we only have one history; using model runs we can generate many histories, each one with completely accurate observations." There is only one real-world history, model-land histories are fictions. It is rather like being in love, having a family, and and making the claim that a room full of novels provides a much richer experience than living one real life.

Will you say that what the model is used for matters?

Yes. When we started trying to fund climateprediction.net over 20 years ago, many, most, climate modellers said there was no place for ensembles in climate modelling, that ensembles belonged only where models were used in weather-like ways. I did not understand this point of view then; I am starting to understand some of its merits of this point of view only now. And when the first climateprediction.net paper was in review, a referee "required" us to present "probabilities," we refused since the model we used could not provide what the reviewer wanted, delaying our paper significantly.

So, what a model is used for matters in at least two ways. If it is a weather-like task or a climate-like task. And if the situation is one in which the model has remarkably high fidelity or one in which it does not.

So, for instance, coming back to something you said earlier about including certain aspects, like buildings or waves, in the computer model, I guess this depends a lot on what the model is for. So, if you have different possible uses, or different possible users, for the model, you want to include certain aspects, whereas for the other one, you would not include; for instance, with buildings, if you are an insurance company, maybe the buildings are very important, whereas if you're a farmer, maybe it's not so important. Does that make sense?

It does make sense, but not as much as many people think. The issue is a mathematical property of nonlinear systems called "mixing." In a system that is mixing, if you get something non-trivially wrong in the short term, then pretty soon you'll get everything wrong. Farmer and insurer, ship captain and energy company: the forecast is likely to be useful to all of them or none of them. It is not so much the application of the model, but the timescale that determines what you need to "include" (to simulate with high fidelity) in a model — how far into the future are you looking to go? Remember a weather model is a simplified climate model.'

So, if it's tomorrow or if it's a year's time…?

So, if it's tomorrow I don't turn on the ocean, because the ocean doesn't change radically in a day and it costs a lot of computer power to simulate even that lack of change; I'll rather put that computer power to get a higher fidelity (more realistic) atmosphere. If it's sometime next week, I still don't want to turn on the ocean, but I can document the advantage of turning on surface waves which makes sense as they actually affect the winds. If it's next year, I need to turn on the ocean; and then all of a sudden I can't spend as much money on my model atmosphere. The farther out I want to look, the more phenomena I need to simulate with high fidelity.

But state-of-the-art computer technology restricts me to a zero-sum game: the further into the future I want to run the model, the more simplifications I must make, and the more phenomena I "should" include. The fact the system is mixing tells me that the more simplifications I make, the sooner the model loses all resemblance to our future.

That's quite interesting. You said that if it's in a week, you need to turn on the waves; so, does it mean that you will have, for instance, six days when you don't turn on the waves, and then on the seventh day you turn on the waves and then you see something happening?

The problem with that approach is first, that we do not know what the waves should look like seven days from now unless they were turned on from the start. Second, that the lack of waves in those first six days means the model-atmosphere is already in a state of greater error. The challenge in operational forecasting is the constraint of fixed total computing power. So, what is usually done is different models are run for different time horizons. One for 6 hours. Another for days to weeks. Another for months. Another for… Brian Hoskins has long argued that the forecast should be seamless, and I agree with him. That suggests getting the best information may require running a suite of models with very different strengths and timescales. I want the forecast to be seamless, but I want a mix of models: my simulations are anything but seamless. After all, if I have waves turned on, I had to turn something else off. I want a mix of models that give me the most information about the future (including the potential conclusion "I do not know.") that brings the intended target back into model construction at a very fundamental level.

So, you mean that to get a very good forecast in seven days, you will get a bad forecast for tomorrow?

That is a great observation. In short, yes: you will degrade the short-term forecast to improve the longer-term forecast, but it makes sense, no? If I have a fixed amount of computer power, then improving the forecast for day seven requires me to degrade the quality of the forecast for day three. For a good probability forecast at seven days, I will use a model which will probably not forecast day two or day three as well as a model focused on what is important to forecast on day two or day three. So, it's a vicious circle. It means that the further I want to go, the more phenomena I need to turn on, and the less I can invest in the things I already know are important. In theory, weather models are simplified climate models. In practice, technological constraints require we turn off things in climate models, things that we know are important, if we insist on running climate models for a hundred years.

What is the aim of the model?

Now that is a really tricky question. For some people (they are often called "users," but once you get to know them, you realise they are really people), some people like those who work in EDF want a good idea of how much natural gas they'll need to heat the south of England the end of next week. They have a very well-defined forecast problem and they have a very asymmetric cost, in that if they do not have enough, things get really bad and very expensive, while if they have too much, their profit dips a bit. Others want to use the model to figure out how much EDF will think it needs and then take a position in the market motivated purely by profit.

Other people, including most physical scientists, actually want to understand reality better, and the operational model provides a well-resourced test bed. With the model they can look inside (model) storms that are so vicious that if they tried to take measurements in its real-world analogue, they would be torn to pieces. For them, the model is key to seeing what might be happening in the real world, for getting cool ideas. For designing and testing real-world experiments that might distinguish between competing ideas (without getting torn apart). Scientifically, I'd rather have a model that gives me good ideas than a model that can predict so and so.

And then there are people (scientists with a naïve realist bend) who see today's model as a step towards the perfect model, and they are pushing to get there. This fundamentally biases their idea of things (like parameter values) away from what is most useful, away from what is empirically successful, and towards what has to be done if we are to ready the perfect model. It is a pity that Paul Teller's "Twilight of the Perfect Model Model" is not required reading in graduate school.

So, we'll come back to the question of good model/bad model, but this is also interesting to see, what you can learn from models which are not extremely well performing on certain tasks.

I agree. One insight can be worth a thousand (good) forecasts. And models which are ineffective as (or never designed to be) forecast models can provide insight and understanding.

So, now the question of computer simulations, which seems to be really the core of the model. The question is about the relationship between the simulation and the model. So, here it seems to be really intricate.

Yes and no. The simulation belongs to the model, like one of ten thousand seeds, it is the offspring of the model. But interpreting information in a nonlinear simulation to better foresee the future, that can be incredibly intricate, one learns a model-based distribution, not the probability distribution of possible outcomes in the real world. I think there is still a lot of improvement to be made there.

By nonlinear…

By nonlinear, I mean things like rabbits.

I didn't expect that!

Let's look back to what Fibonacci suggested back in 1202. Put a pair of rabbits in a garden and then a month later I've got two pairs of rabbits, and then a month later I've got three and then five, and then eight, right? And this grows exponentially. So, by now rabbits should have taken over the world. In linear systems, the response is proportional to the change, in nonlinear systems, this need not be the case. The increase is bigger at each time step. And the fact that rabbits have not quite taken over the world over the last 800 years suggests there are further nonlinearites, negative feedbacks, say, due to the finite land area on the Earth, that must also come into play.

That's perfect. And what is a good model?

What constitutes a good model is without doubt in the eye of the beholder. A good model gives you insight you did not have before. I think it is pretty hard to get beyond "One makes better decisions with this model than without it." (And afterwards, there is no justification for the view, you were "just lucky.") For an academic, a great model reveals something you did not know, gives you a cool idea, illustrates something you believed but could not substantiate in detail, tells you what to measure next. Good models capture, illustrate, compartmentalise the things you do know; often things we know well, but cannot work out the implications of on the blackboard because it just gets too complicated. If a model gives horrible forecasts, but in doing so reveals a better understanding of the relevant physical processes, if it teaches you what to do to build a better model, would you call it a bad model?

The Lorenz Equations and the Moore–Spiegel Equations do both. They illustrate what it means for simple equations to yield chaotic dynamics. They show that chaotic systems can have periods of time when it's easy to forecast, long periods where uncertainties actually shrink with time (chaos merely requires this doesn't happen forever). I did not see this coming and it was a surprise to many other people also (although I think Ed had noticed it). Ed Lorenz created those equations because he wanted to show that the Wigner Filter might well be mathematically optimal in particular applications, but that it was far from optimal in meteorological forecasting. So, his 1963 model was "good enough" to do that; finding chaos was a bonus. Moore and Spiegel wanted to show that the complicated oscillation observed in stars did not require either complicated physics or randomness. Their 1966 model did this, finding chaos played a central role in making their claim.

What makes a good model really depends on what counts as good for you: what are you looking for, are you looking to take action in the real world? Are you just trying to understand it better?

On the other hand, if you're trying to make a forecast, a good model might give you a probability distribution on what the temperature for the end of next week in the southern half of England will be, and you might really want that for various reasons; or you might just want to know is there more than a 10% chance it will be cold; or you might even only want to know is there a chance there'll be a hurricane. When I was a kid, no Floridian meteorologist would ever say, 'there's no chance of having hurricane force winds tomorrow'. It is just a very dangerous statement to make.

What has been the impact of the development of new technologies or tools in your field?

Huge. Arguably ungaugeable. And technology remains a limiting factor for weather, for seasonal and for climate forecasting. In about 1917, during the First World War, LF Richardson computed the first weather forecast by hand, while he was an ambulance driver in France, and it took him two-and-a-half years; it was lost once then rediscovered under a pile of coal. And: it was wrong. Arguably, the model was still "good" because he understood why it was wrong. He never intended to compute the forecast before the events came to pass. Rather, similar equations were used in the 1940s and 1950s, in part to test and better understand mathematics on a computer, in part to inform weather forecasters. Weather forecasting was developed because it was a hard, useful, interesting thing to do that people could talk about; computers were also used for other things that people couldn't talk about. So weather was in some even broader sense; a model itself. Richardson actually said that we'd need 64,000 computers — meaning graduate students — who would hand pieces of paper back and forth, left and right, to accomplish computations fast enough to get a forecast before Nature showed us the result. Basic technological limitations have always been with us. Three decades later, Charney, Van Neumann, Phillips, and their guys actually executed this program, having learned from Richardson's work. Another two decades and computer models were outperforming human weather forecasts routinely in the short range. If you talk to an applied mathematician about the size of the computer required for a fair chance of an acceptable-fidelity first–principles simulation of the Earth's weather, they say the computers aren't quite big enough just yet. And climate requires much more. Weather models are simplified climate models. So on the same hardware one can expect better weather simulations than climate simulations. Technological constraints force us to omit things we know are important in a climate model if we insist on making climate runs for a hundred(s) years. The IPCC states explicitly that the information in these runs drops significantly after 2050 even on global length scales. Given the targets of interest, I believe that there are more useful ways to deploy the computer power dedicated to climate simulation.

Another aspect of technology is communication in weather and in cosmology — it's not just having the telescope, or the satellites, to look at what the Earth is doing, but actually being able to send that information to Earth fast enough that the astronomer doesn't die before she has the data to evaluate the claims in her proposed thesis. Weather information from satellites is already so voluminous, most of it is not used in real-time weather forecasts. Both communications links and computer power set critical barriers to improving real-time weather forecasts.

Just a remark here as I wonder whether this applies to your field or not: when I spoke to Gerd Folkers about drug design and pharmacology, it came out that computers are extremely important, because you do the simulation instead of using labs. Then we had a very interesting discussion about moving from science to engineering, and somehow having

more and more computer power limiting the thinking process — because it's so easy to program something rather than maybe having to do things by hand and thinking about it carefully before doing lab testing. So, I was wondering whether you feel there is a danger here in your field as well, or this is not relevant?

I think there is a huge danger, but I think it's an artificial danger. I think it is often possible to do serious simulation, but it's faster and cheaper to do sloppy simulations. As an undergraduate, I had to punch a deck of cards and wait a while for each run of my model; I was heavily incentivised to code carefully. Being able to simply tweak code until it compiles can easily lead to sloppy simulation. We need closer links between simulation and experiments (Folker's lab tests). And we need to honestly report when "we don't know." When we believe the simulation is not adequate to answer a target question in the real world, one should never couch doubt by reporting the model output as "the best available" or quantifying what "the simulation says" in model land. The question is: is the output thought adequate for purpose? If it is merely "Best available" and thought not to be adequate, the model should play a very limited role in decision making. I worked with Team New Zealand on one of their Americas Cup racing yachts, and what they would do is first do simulations and then they would build a test hull about as big as my office, and they would try it out in the water tank. And then they would do a bunch of simulations and they would build another one. They would build three test hulls; and then they would build a full size hull, and they'd try that out in a lake, and then do more simulations. So, constantly going back between simulations and scale models, and then finally real models. And in the end, they built three full size boats and sailed two of them. The key is maintaining that connection. And very often, it's so easy to do what I would call 'looks like physics' — you build a computer simulation that looks like your target system when viewed from the right angle, but you don't even understand how it works; as opposed to really trying to do serious simulation, and sometimes failing robustly (and admitting this failure). This is also critical and very often there are mathematical symmetries — you know, the Earth is sort of round, and there's a sense in which you can save a lot of computer time by designing your simulations to take advantage of certain aspects of the equations themselves, or you can just slap it all on the computer. There are advantages to both ways, but if you just slap them on the computer, at the end you need to check that the solutions still obey the symmetries in your original equation. And if they don't, you know without looking that something is fundamentally wrong. I think the real danger is that people do these 'look like' simulations where they may not even record what they do. Economists are famous for having taken the wrong column from out of an Excel spreadsheet and changing government policy. This idea that you're not seriously focused on things like how you design the model, this is a significant challenge to simulation and prediction. On the other hand, the models in drug design, they allow you to test things billions of times more complicated than you could test, and billions of times more things than you could test in the lab. The key is that you just want to make sure that the tests are worth doing. So, I share the concern, I think it's a huge problem in all of applied simulation; but it's really one of sloppiness. If you look at people who either love what they do, like Team New Zealand, or you look at people who are really concerned about safety, like nuclear stewardship or aircraft design, these people write tight code. Other things which have equally big impact on humanity, be it drugs or climate or finance, there's an issue that I certainly share with Gerd Folkers.

Going back to something we mentioned earlier briefly — uncertainty. How would you define uncertainty?

People have so many different internal definitions of uncertainty that I try to avoid the word. (This is actually quite fun as a game, to challenge all the lecturers in, say, a summer school to avoid using the word "uncertainty," saying instead what they really mean: Imprecision? Ambiguity? Intractability? Indeterminism?...). Nick Stern and I wrote a paper on scientific uncertainty and climate policy which has five or so concepts, each called "uncertainty." I witnessed a discussion between a former governor of the Bank of England and a former president of the Royal Society in which one was talking about Knightian risk (which physical scientists tend to call uncertainty) and the other was talking about Knightian uncertainty (which physical scientists tend to call ambiguity, or "deep uncertainty," the case when you cannot assign a relevant probability). Sadly, I did not realise this until the conversation was almost over; it was much too late to jump in and try and clarify things. I expect that the word "uncertainty" has caused more confusion in interdisciplinary discussion than almost any other word, it is up there with "optimal" and "true." As a physicist, when I say uncertainty, I almost always mean "imprecision": there is a precise number out there, I just do not know exactly what it is. In this sense, Heisenberg's Principle is one of indeterminacy, not one of uncertainty. Uncertainty takes us back to the number of rocks in my office. There is a well-defined integer that is the number of rocks in this room, I just don't know whether it is 32, but is near 32-ish, I am pretty sure there are more than 16 and less than 64 rocks in my collection. Uncertainty as impression (the number of rocks in this room) is rather different from uncertainty as indefiniteness (the position and momentum of an electron are not defined at the same time). And these are different again from the uncertainty in a calculation which we know how to do but which is intractable today, but will almost surely be tractable in 2048.

And where does uncertainty play a role in the modelling in your field, and how does modelling help you to understand uncertainty?

I am interested in chaotic systems, or better said: I am interested in physical systems that are best simulated with nonlinear models which appear to be chaotic. In chaotic systems, uncertainty does really interesting things. And in any system, being able to track the imprecision of your knowledge of the present into imprecision of your knowledge of the future is a valuable talent. Models aim to allow us to do that, to track uncertainty due to imprecision in the initial condition as we move into the future. Just as your models aim to allow you to track the creation of uncertainty in your stochastic models as one moves into the future, models can help us understand the evolution of uncertainty. Even to better understand which kind of uncertainty is most relevant to decision-makers: yours, mine, or "other."

Do you mean uncertainty in the inputs?

Perhaps. In truth, I am most interested in whichever type of "uncertainty" leads to the most confusion. But let's say there is uncertainty in the inputs. Like the number of rocks in this room. There is an answer, and it is an integer, but we don't exactly know what it is, we can put a probability on it. When we're dealing with this kind of imprecision, I'm a Bayesian. And in this case, I even think all Bayesians agree!

In other situations, we are forced to face ambiguity. Where we have been discussing things we think are precisely defined, but for which we do not have enough information regarding the targets to define a probability, we would be willing to treat as such. Ambiguity.

One can always redefine probability so that you could always claim it is well defined, of course. Such people are rarely willing to take bets based on their probabilities (at least not for very long). Such maneuverers are easily shown to make the probabilities provided all but irrelevant for decision-making.

There are things we don't know enough about to put a probability on — they're either non-sequiturs, they don't match, or we just don't have the information to put the kind of probability distribution on that anyone would want to put a bet on. You could put one on anyway, but that's a different issue. So that's Knightian uncertainty, physical scientists generally call it ambiguity. Then there are these issues of tractability, right: we know a lot, but as soon as — what Laplace worried about — as soon as you do a calculation, you introduce approximations. And these different kinds of uncertainties really play very different roles. We have all of them, as the measurements we make aren't precise — we have multiple thermometers at Heathrow, but they don't all say the same thing. We actually don't really know the correspondence between temperature at Heathrow as we measure it and temperature in the model point that's closest to Heathrow. And we know that the equations in our model, have model parameters which may or may not reflect an actual physical value. If the model-parameter is undefined in the real world then what does it mean to say its value is uncertain? It is undefined. It is indefinite. It is not imprecisely known in these equations... Let's go back to one of Newton's Laws: it just has one parameter — big G; there's only one constant, one constant of Nature, and this was a constant in Nature for a long time — Newton's universal constant of gravitation. But it was uncertain — we only knew it to a few digits ($\sim 6.6743 \times 10^{-11}$ N·m^2/kg^2) number, for your chemist. But now it turns out that Newton's Laws are wrong — we need general relativity. So it's not that big G is uncertain, it's not imprecise, it's undefined, it's indefinite, it's indeterminate. Like a quantitative property of the luminiferous ether, it's not that we don't know what value to use, there *isn't* a value to know.

In such cases I'm not a Bayesian. We can ask what value should we use in the model — the best value; but that depends on the question; and an aim of physics is that we get away from that. So, this universal constant, there's still some out there, but we don't know any of them with the sort of precision that a mathematician would want to say for this incredible utility of mathematics.

So, it seems, as you said earlier, that you have different types of uncertainty. So for some of these types of uncertainty, do you have some sort of sensitivity analysis or robustness analysis?

Sure, we do exactly that kind of thing. Of course, it is not immediately clear what it tells you beyond the fact that your model is sensitive to small changes. We test different initial conditions and see how quickly our ensemble of simulations spreads out as it evolves into the future. In a similar way, we can test a distribution of different values of each parameter in the model. If we know that our model has never been able to shadow for some parameter values, we might discard those values. And we often find that our model is very sensitive to certain initial observations; happily, we can use that information to improve the forecasts, sometimes.

How can you do that?

So, for one summer there was a plane based in Reykjavik and a plane based in Norfolk, and on 14 days the weather service could tell them to switch places and dropping little devices out of a window of the plane to measure things like temperature and humidity before they splashed into the Atlantic. The question is: what days do you tell them to go? and where do you ask them to fly? The answers came from a simple sensitivity analysis of the aspects of the weather over the Atlantic Ocean of which we don't normally observe very well, and it appears there is a higher than normal chance of a big storm coming for Europe or to England? So, if we could figure out those days when the chance of a storm is high, we could have them fly on those days and take measurements which would most reduce our uncertainty on the initial conditions. Well, not the absolute uncertainty, but those uncertainties which would most reduce the uncertainty in the forecast.

Well, no, wait: doing this may make the forecast, the ensemble of simulations, tighter, but we still have the additional question of whether the sensitivity of today's model yields better forecast reliability.

That makes sense. You have uncertainty in the inputs, and uncertainty in the parameters. Is that everything?

No. There are two more we know about. Perhaps the most wicked uncertainty is in the mathematical structure of the model itself. Model Inadequacy. Since different weather centres have different numerical models, which make different approximations, maybe to the same theory; maybe they even have slightly different theories. So, another thing you can try to sample is different *kinds* of models. And this is *very* different.

The initial condition may lie in a 10^7 dimensional vector, but it's still just a collection of real numbers. We can sample those. And we can put distributions on parameters. But how does one sample the space of all good models given our understanding of science today?!? And even if we could, what would that ensemble of model-simulations tell us?

And these models are different by nature? I mean, let's try to simplify if we can a little bit. So, you would have models with parameters, variables, and as input, you run these things and you get an output. But they would be different relationships, or different by nature, these models?

The models are different by nature. Two models might be aiming at the same mathematical target, but each approximating it in a different way, while a third might be aiming at a different mathematical target. And then, given the computer code, the particular computer (or compiler or …) you use can make a difference.

Really! How?

Each year, I give the students in my statistics class the same simple equation, the Logistic Map we defined above. I ask them to iterate it, to move it forward in time 64 steps, starting with the square root of two over two. There are about 20 students and usually 12 or so different answers. Of course, if they were using the same computer, and they had written exactly the same computer code, they would get the same answer (assuming no cosmic ray hit the computer's memory chip); but just because they put the parentheses in different places, or used different computers and mobile phones, the digital trajectory is different. They are sampling a function space, which is really rather different from the real numbers. Once they see they all have different answers, I can ask them what the probability is that 64 iterations of the square-root of two over two is greater than a half?

Some say 50/50, since they know they do not know. Some report the relative frequency with which their simulations were greater than a half. But this is, of course, a well-defined mathematical question. The probability is either zero or it is one in the case of a mathematical statement. Are things really any different when forecasting the weather? Or the planets? Or in finance?

So, for model uncertainty, is there any way that people do some blending on these models in order to take into account everything?

I'd argue that the fact that we can never know whether or not we have "taken everything into account" and that requires us to state, quantitatively, what we believe the probability is that we have left out something important. The probability of a "Big Surprise."

And it is not (only) about "unknown unknowns." While I think an intuitive scientist or decision-maker can have valuable feelings about the trustworthiness, often these come from "known neglecteds," things which we know are likely to play a role, perhaps a major role, in the outcome, but which the technology of the day does not allow us to include. My model has to run on 2016 hardware; and a prediction model has to run faster than real time. Scientists often have a gut feeling for how far into the future we believe the model to be informative, or at least at what point in the future we no longer believe the model to be informative. This in part is due to unknown unknowns, but also to known neglecteds.

In the situation where I believe everything that is important is in the model, so that the model will produce high fidelity simulations, then the "Probability of a Big Surprise" comes from the chance of learning something really new. It is almost win–win, scientifically. In that case, I would say the probabilities I extract from my simulations are mature. I do not expect them to change unless I learn something new, new observations or new ideas. On the other hand, if I expect the Big Surprise to come from having omitted something for practical reasons, or technical reasons or computational constraints…in short, from a known neglected, then if I provide any probability distributions to decision-makers, it is critical I also provide my Probability of a Big Surprise. I never want to be in a position of claiming plausible deniability based on the fine print in my report.

Does blending the models account for this chance of a big surprise?

I'd say it can reduce some sources of surprises, but not those due either to shared ignorance or restrictions from state-of-the-art computation. It is clear that the diversity of our models is not a realistic measure of the probability in the future, but blending is likely to increase the probability on the outcome, unless one of the models is perfect. (And none of our models is perfect.) If you look around the world at, say, different weather models in different national meteorological offices, they are each designed to be the best they can be; they were not designed to complement each other. So, what we have is an ensemble of opportunities, could we construct a more complimentary, more informative ensemble. And, of course, they all share today's unknown unknowns.

So, risk now: how do you define risk?

As always, I am happy to use one definition in any one conversation of risk, as long as everyone is willing to take care and stick to that definition. Note that to most physicist's uncertainty means imprecision, while to most economist it means Knightian-risk. But given my choice, I'd define risk as a much more personal thing.

There is an insightful book called *The Feeling of Risk* in which the author admits that despite years of lecturing on risk, he failed to feel, to really feel risk until the day his car broke down on the central embankment of a super highway, and he had to run across six lanes of high speed traffic. In the middle of this, he realised "Hey, ***this*** is risk." (Happily, the insight did not distract him; he lived to tell that tale.) I personally like to take long walks on my own in the wilderness, in the high desert of New Mexico, on barrier islands off the Florida–Georgia coast… There is a risk that is related to the chance things that might happen during a hike; there's a reward that somehow makes the risk worth taking, or not. I don't estimate the probability some "bad thing" will happen. Rather, I ask myself how much that piece of gear weighs, might I use it for something else? How bad would it be if that "bad thing" happened? Do I really want to expose myself to that risk, with or without this piece of kit?

The real risk, the risk you feel, is much more conceptual, more imaginative, scoping counterfactual possibilities without being distracted by them. There is some chance that there will be a small rockslide and I break my leg; or am bitten by a pit viper. Some Bayesian might be able to put a probability on that, or work out the probability I appear to put on that. But I have thought through all those possibilities and am eager to go without postponing my departure to do calculations. The risk I feel is not from the probability of an event that I have realised might happen, feel prepared (equipped) to face should it happen, but rather the risk I feel is from the unknown things I cannot on reflect on. Those things are not included in the Bayesian calculation, except perhaps by including the probability of a "Big Surprise." The things I do not even realise might happen. Risk is embodied in the situations that I'm not prepared to deal with.

Or in finance: there is a thrill in forecasting what a market will do today before they open, advising traders, and then seeing how the day unfolds. I'd argue risk is a different feeling: Risk is what you feel when you are suddenly asked either to buy or to sell 100,000 barrels of Brent Sweet Crude, then handed a phone as someone hits a speed-dialer. Risk takes on a much deeper, richer feel when you are part of the system, not an analyst.

So, like many of the forecast users that I work with, risk for me is not so much a technical term; it really is more conceptual much more immediate. It's just not some event in my head that I can put an uncertainty on. Risk regards something I have not prepared to deal with. That I am surprised by it, astounded by it; yet understand it (after the fact). Not something I can blame on "bad luck" or a low probability event happening;

rather, insight such that I would have made different decisions had I had that insight before the event.

So, would you say that uncertainty is something which is more objective and risk is more subjective?

These two words are used so inconsistently I would avoid them. But as you asked I would say that imprecision is objective and quantifiable when we believe there is a true value which we know only imprecisely. Ambiguity is objective, but more difficult to quantify, perhaps impossible to define. Risk, for me personally, has more to do with being comfortable with a vast number of things that I've got no quantitative handle on at all, along with my accepting that there is a good chance that something completely unexpected (by me) will happen. What goes into this category is, of course, heavily influenced by my personal/individual experiences.

And the last question, which is very personal as well: for you, what is the experience or result in your field — so it can be one of your own or somebody else's — that has had the most significant impact and why?

I fear I can only reduce it to two; but one is very short. The first was a remark by my PhD supervisor, Ed Spiegel, who took me into a room full of accomplished scientists, it was Walsh Cottage (at Woods Hole Oceanographic Institution) and said something like "watch these guys, figure out who the serious ones are, then simply do what they do." It took a while before I realised that this is really the only formula there is for doing good applied maths: you have to think, and it helps a lot to have to explain your work to a room full of smart people with completely different backgrounds. It leads to a good answer when there is one, and an interestingly robust reply when there is not. It is a different approach to science than now common turn-the-handle-and-get-an-answer, make-a-picture, publish-it approach. The old approach leads to lots and lots of arguments — there are huge egos in a situation like at Walsh Cottage — but there are basic ground rules about what counts, what is accepted as evidence; and... it's quite honourable — they might cheat if they could, but they can't. If the argument doesn't stand up, forever, they failed. Perhaps this even leads back to the issue of how you program a computer: is the code OK when it compiles and runs? When it gets test problem "right"? When do you trust your code? (Answer: no good programmer ever trusts their code.) How seriously you take the simulation that your protein-folding friend is worried about? Good modellers are probably rather like good bench chemists, they are over-careful. After a few additional years of observation, my favourite living Baysian, Jim Berger, expected there was a bug in the code he used to make hazard maps for the volcano mount; they found the bug, corrected it, and made new maps. Does he believe that was the last bug? No. What is a real world Bayesian to do?

The second big challenge was just when I realised that, in fact, the uncertainties we understand are not the problem; we understand chaos, it can be costly to cope with chaos, but we know how to do so. I spent a lot of time doing nonlinear dynamics of chaos — in a very real sense Kevin Judd and I solved the prediction problem (for chaos) in the late 1990s, while one can only make probability forecasts, you can make them as sharp as you like; it's just a question of how much you want to pay for. And then, we spent 10 years trying to find the physical system where chaos *per se* was the limit to predictability. Chaos was never the problem. It's not imprecision that's the problem, it is the fact we have the wrong equations (model inadequacy). When you really look carefully, it turns out (so far) that our models can't shadow reality to within the observational noise, casting a shadow on the so-called "incredible effectiveness of mathematics." From Newton's Laws of Universal Gravitation to Kirchoff's Laws for an electric circuit, the dynamics of the model "looks like" that of a circuit, but the equations cannot shadow <u>that</u> particular circuit, or the actual thermally driven rotating fluid annulus in Peter Read's Oxford lab, the atmosphere, or the solar system, or…

So, the real insight was that our models are useful, but there's not an isomorphism between the model and reality. And there's probably not an isomorphism between the theory and reality. So then, what's the difference between doing science and doing engineering? I think there still is one, a big one, but it would help if we learned how to draw a clearer line between mathematical analysis and analysing actual dynamical systems.

Feynman said there were two possible end-games for physics, and that we could not know which one would play out. Either we'd reach a theory so precise and beautiful that we'd know that was the way the Universe was (not that we could ever prove it was true), or we'd forever generate more theory, what I'd call "approximations all the way down." He hoped it would prove to be the first, and in that case, mathematics truly is unreasonably effective.

Many of my pure mathematician friends in Oxford are Platonists, they really believe there is a mathematical description of the world, but it doesn't affect anything they do because they work wholly within that mathematical fiction. On the other hand, people who interact with the world, people who forecast real things, often aim to get insights from the models about how things evolve into the future, or even how the system will behave when you change this or that. Applied mathematicians see their work in action.

In practice, the role of model inadequacy is vastly undersung. It is somewhat challenging to write a viable grant proposal that states clearly that fundamentally we do not know how the system works and will be in the same position at the end of the grant. It is very hard to teach and harder to examine: it is much easier to write well-posed mathematics questions on how to propagate imprecision, compute linear sensitivity, work with precise probabilities. But in practice, such precise manipulations rarely reflect reality; realising that it wasn't that I was solving the problem incorrectly, but that I was attacking the wrong problem. Realising that the challenge was not in mathematical manipulation, but due to Model Inadequacy given the targets we aim for and our interpretation of the results. That was a big change.

Chapter 12
A Conversation with Michael Stumpf

Can you describe your field briefly?

My field tries to understand biological processes quantitatively and predictively; in the narrower sense, what I'm trying to do is to understand how we can learn the structure and dynamics of biological systems from data.

So, can you give us an example of what a biological system is?

A biological system could, for example, be a cell — either a bacterial cell or a cell in the human body, for example, a stem cell — and it has certain, what we call, decision-making processes that would allow the cell to respond dynamically to a changing environment. In the case of a stem cell, it could, for example, lead to the stem cell dividing and giving rise to more differentiated cells, which then make up tissues such as blood and skin.

Can you give us an example of a particular model in your field?

We often use what's called pathways or gene regulatory networks; these are descriptions that tell us which gene activates with which other set of genes. And for these types of processes, we develop mathematical models — either differential equations or stochastic process descriptions of how they might alter their behaviour over time and affect the dynamics of molecules over time and so on.

Could you give us some more details on various steps involved in a typical modelling process (for instance, the stem cell example)?

In the first place, we learn from our collaborators, who are domain experts. This, in addition to reading the literature (again potentially guided by what our colleagues suggest), leads us to develop a simple scaffold around which we model. Here, it can sometimes be that we start from something simple and then extend it bit by bit until we have something that seems to serve its purpose. Or, we start from something fairly complex and pare it down to the point where comparisons with data become possible and meaningful. Once we have such a model, we can try to learn the parameters; then we try to assess formally, as much as possible, how robust our analyses are to uncertainties in the model: that is, would we get an equally (or nearly equally) good agreement between models and data for other models than the one we developed.

We will come back to the role of mathematics a bit later. If you look at these stem cells, could you explain why a model would be needed?

Stem cells are actually quite a nice example because a stem cell is a cell which, if you take an embryonic stem cell, can develop into every type of tissue that you have. So, what we want to identify is what is the process of going from this undifferentiated cell to something which could become a hair or part of your retina and so on? People have come up with models that look at a cell being like a ball in a valley, and then one valley would be the stem cell, another neighbouring valley might be a more differentiated cell, and finally down the last valley would be a retina cell. So, this is a model which tries to map the behaviour of stem cells and their offspring — it's a pictorial representation of what drives these processes and describes them.

I see the role of mathematics in the modelling. Would you say that mathematics in modelling in your field is essential, or is a tool?

It is one aspect of studying Nature. I think it is at least equivalent to other techniques that biologists have been using for a long time, such as X-ray crystallography, stuff that they use to deduce the structure of DNA, or of proteins — electron microscopy, microscopy, and so on. A particular advantage that mathematics has is it can elucidate stuff which you can't directly observe. So, you know that certain things are related somehow, and that can be literally related — so, we know that we are related to chimpanzees, and because we have this model that we're related to chimpanzees, we haven't observed this split between chimps and us, but because we know we're related we can then take genetic diversity in humans and chimps and we can say that the split between our ancestors and chimps probably happened about 5 million years ago; it's something which is hidden, so the family tree between us and chimps is hidden, but it's a structure, and because we know the structure exists, we can start with its mathematical descriptions. I hope that makes sense.

Just for me to be very clear, the role of mathematics for you could be in two places: either because it's impossible to observe, and it will almost always be impossible to observe, so it's a way to represent something we cannot observe; but can it be also, let's say, before trying to figure out how to observe and what sort of experiment to do, you have a sort of mathematical formulation that you want to test? Am I correct?

We have done this, we have spent a lot of effort on experimental design — what experiments biologists should conduct to learn more about the inner workings of a cell or of an organism. So, that would start from us having a candidate model — a starting point (we can discuss later where these starting models come from), and then try to see if we want to understand how closely that model represents the reality we have to do experiments on. Not all experiments are equally informative, and we can, using the assumptions of this model, try to see what experiments would allow us to get the best agreement between model and data, or allow us to rule out a model.

So, how does this model that you would like to verify or to test arise? Does it arise from previous experiments or something else?

For lack of a better word, it is a diffusive and creative process, and typically it requires a biological expert to sit down together with us if it's something where we haven't got that much experience, although in many cases we do know a lot about the biology ourselves. We would take what has been published in the literature, what are widely held beliefs, and we have very good insights about, for instance, how different proteins interact — so, for example, here we have a protein [shows model] which consists of six identical units, and we know that these six identical proteins bind together and only when they've bound together can they fulfil their function. And many of these processes we do know, and we can then write down equations that correspond to these processes.

Now going back to the question, because it's really interesting, I think it's the first interview we have done where experiments are really an important feature, so understanding how experimentation and modelling go together is quite important. So, that brings us to the next question which is what constituents besides a mathematical formalism are part of the model? It seems that experimentation is quite an important part?

So, for us, the experimentation is not part of the model, but experimentation is crucial to fine tune or refute aspects of our model. So, I think, for us, a model without direct experimental relevance would be pointless. We wouldn't do this.

You mentioned that for a model, you have certain assumptions, and then you have what links together the various quantities and then you will have the output of your model. Is experimentation more about testing whether the assumptions you're making are right, could be satisfied, are plausible, or testing about the output?

That's quite an interesting question because in our field you have two different types of model applications. In the first one, you want to have mechanistic understanding — how do these things work? In the second one, you want to increase your predictive power — you want to get better predictions. And there's a trade-off between these two things. The model that is best able to predict future outcomes is often the one where you have very little mechanistic information, and that's for reasons which go into very deep mathematics and statistics. We do both, but we are principally interested in understanding how Nature works, they don't want to predict something if it doesn't coincide with getting increased knowledge. So, the experiments that we are most interested in doing, they test the assumptions.

Can you expand a bit on the link between assumption and model?

The two are, for me at least, inextricably linked: the model should provide as mathematically or computationally a representation of our assumptions as possible.

How important is notation in modelling?

Notation should help and clarify things. But it must not become an aim in its own right. There are probably people who disagree with this, and clearly there are some criteria that we can demand our notation to meet.

What is the role of language in modelling? Or any qualitative aspect?

The problem with that is that the meaning that you attribute to words tends to change the longer you spend thinking about things in a certain way. So, for example, the word 'network' might mean something completely different to me than it would mean to someone who is in charge of running the London underground network. And the more I've worked on networks over the past 10 or so years, the more my understanding of the word 'network' — or the particular meaning that I have come to associate with it — has become more natural to me. So, I've got a sort of intuitive understanding of what *I* mean by network, or how I view networks, that is very different to what it was 10 or 12 years ago. So, language has an important role in the sense that it can often cause confusion. So, if I use a certain term to which one of my colleagues is not accustomed, then that can cause confusion. So, in a way, the mathematics is a more neutral way of doing it. Trying to interpret the mathematics together with our experimental or clinical colleagues, we have to find a common language, and that is a problem. It's not a problem with the modelling, but with how you view the world or the relevance of the model to the world.

Yes, I think it's something we noticed in other interviews, and this is also the problem of disseminating research and results — there is this problem of putting quantitative things into qualitative terms.

I think that is more of a problem if you have to deal, for example, with the direct applications of the research. So, for example, one of the projects that we're interested in is trying to design bacteria that live in the soil, close to the roots of plants, and that help plants to absorb nitrogen better. The advantage of such a bacterium would be that you can use much less fertiliser, but it would be a genetically modified bacterium, and you would have to explain to the public and also to regulators, why this is a good thing. And there, the language that you use can be highly incendiary — we have to be really careful. That's not the problem that I mentioned before, how I would view a word or a model differently.

Coming back to models: models are often said to represent a target system, say, a selected aspect of the world. Does this characterisation of a model apply to your field?

Yes, it applies in one sense: in systems biology, there have been broadly two schools of thought — they go by the name 'top-down' or 'bottom-up' modelling. In the bottom-up approach you would take a small set of variables, of states that you want to describe, and you would get fairly detailed mathematical models of those things. In the top-down approach, you would try to take lots and lots of data and try to get some sort of relationships between the different data points that you have in your dataset. And the latter approach tries to learn model structure from the data; in the former approach, we try to take a small set of reality and try to describe it using a mathematical model. And there, this use that you outlined is indeed how we view models. So, we would take an aspect of reality, you would try to take all the different players that contribute to or explain this aspect of reality.

For instance, in the example that you gave earlier about the bacteria close to the roots of the plants…

Yes, it would be different molecules inside the bacterium that determine how the bacterium digests and releases nitrogen to its environment. We would take these different molecules and we would try to see how they interact with one another, how they change their behaviour as the nitrogen concentration in the environment is altered and so on.

How would you describe the interface between a model and the real world?

Well, it really depends what you want to use the model for. If you want to use the model as an abstraction of reality, then you want to identify the minimum set of variables that would explain the particular target that you are after. The question then is to determine to what extent is that abstraction the only one that could explain the data, and to what extent does your abstraction bias your future analysis of the system. That's something we've invested a fair bit of time in, trying to see how many similarly complex or simple systems can explain real-world data.

Can several models co-exist to look at the same reality?

If you think of reality being available to us only as experimental data, and we have a finite amount of data, we find that many models are equally powerful at explaining reality. We can then try to design better experiments that allow us to distinguish which of those models can be ruled out, but that's the process of research. Just in late December we published a paper which showed that often the number of models that can explain even very detailed data can be very, very large…

And very different from each other?

Yes. So, if we take a model which we know is true because it's a simulation model, you simulate data from that model, and then you try to see how many other models could also explain that data; in one case, I think it was something like we looked at 40 million potential models, and we found that on to the order of 10,000 or so of those models were as likely to explain the data as the true model is. So, to the order of 10,000, it might be 10- or 40,000.

240 *Dialogues Around Models and Uncertainty*

Does this huge number of models differ by the type of assumptions, the type of equations or relationships?

It's the type of relationships. We allow for different relationships between similar molecules. It's like a model that models how populations of molecules inside a cell would change over time, in response to an environmental change, and then we produce similar [simulated] data and we try to see how many other models explain the data.

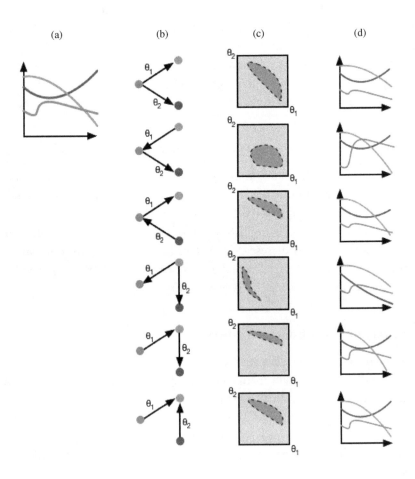

When you say you can whittle it down to say 10- or 40,000 models, if there's a case where you can whittle it down to much fewer models, are they more useful, or is that not the case?

Nobody knows. We've got figures; I can show you if you want to have figures in your piece?

Yes, if you think having figures (in black and white) or a graph can help the understanding.

This is a good figure. From the simulated data, we know what the true model is by definition. So, the true model is, if you have some experimental noise attached to it as well as your observations and we always have some noise, something like the 10,000th best model — so, there are 10,000 better models in terms of fit of model to the data than the true model, so arguably these are the 10,000 models that would be better models, and there is no *a priori* way for us to tell how many of the models are useful or good models to follow up on. But what we can do, is we can try to see what the majority of all the models that can explain the data have in common. Then we'd say, well this feature, the interaction from A to B, is only represented in 400 out of these 10,000 models, whereas the interaction from B to A, the reverse direction, 9,000 have this interaction, so we would then bet money, 9,000 to 400 on interaction B to A rather than A to B. So, this can be used, and there's a very nicely developed statistical theory as to why this might be the case, why this is a good way of doing it.

I have lots of questions related to this! You mentioned better models. What for you would be a good model?

Well, that is a highly personal question in a way because people have different demands from the models that they investigate. I want to have a model that explains how a biological process functions best, that takes into account all the information, but is capable of making non-trivial predictions that are then borne out by further experiments. So, I want to get mechanistic insights into how a stem cell or a bacterium works. There are other circumstances where we want to have a model that best predicts future outcomes. For example, in a spam filter in email, you don't care if the spam filter understands the motivation of the person sending you the spam, but you just have certain characteristics that with high fidelity can weed out the spam — so, that's a purely pattern-matching statistical model. And that's something that has its value especially in clinical predictions if you want to say 'will this person respond to that drug, yes or no?' Then you don't really care if you understand the mechanisms as long as you make the correct prediction. And there you have to make the trade-off, and we try to do both in different circumstances, but if you have to ask me what I want to do for the rest of my working life, I would rather do the mechanistic insights.

You mentioned earlier about partial differential equations and stochastic processes, so we are talking a little bit about different types of relationships, more deterministic ones and introducing randomness because we are moving slowly towards simulations. How would randomness play a role in your model?

We always try to use the modelling approach that is best suited for a given scenario that we are investigating. So, the people in our group have experience in all these different modelling approaches, and whenever we need to, we take randomness into account, but there are three levels at which randomness can enter into our work: first, and in some ways the most trivial, is that the observations we have are subject to noise, and that's one level of randomness; the second is, there is a level of randomness due to intrinsic stochasticity, so many of the reactions that we're looking at are best understood as stochastic random reactions, and if we have to do this, then we'll model this; and there's a third type of randomness which deals with the uncertainty we have about models, so we are trying to model systems where we can't measure everything — so sometimes, if we have a parameter which tells us how quickly a molecule is turned over, then this will depend on the availability of other molecules inside the cell. For example, there's one type of molecule called the ribosome which is the production engine for proteins, where proteins are produced, and the number of ribosomes varies between cells, and we often model this as a random variable — we often model the connecting parameters of our differential equations as random variables. So, that's another level of randomness — that's a highly technical level of randomness, and I'm not quite sure how that can be distilled down further.

We will come back to uncertainty in a moment. What is the relationship between a model and theory?

Well, I think in a way it's a personal use of language, but we are doing theoretical work, and we look at models, and models are always specific to a given scenario that we're investigating.

What do you mean by "scenario"?

A scenario could be bacteria, stem cells, a particular scientific problem. So, we could use models for each of those problems. There is an underlying theory of stochastic processes, there is a theory of evolution that governs everything we're doing, and there are certain types of theories, but they would have a higher level status for me than a model has.

Would you say, for instance, that an expert or researcher in another field would be able to rely on the same theories and develop models for completely different, if I use your language, scenarios? For instance, in physics, maybe they will use similar theories, maybe not the theory of evolution, but some common set of theories?

There is a common set in terms of thermodynamics, for example. My background is in theoretical physics, and there is much more reliance on first principles of very fundamental theories in physics.

Which you don't have in biology?

We don't have this.

Is it because — excuse the non-expert comment — it is a bit like people working in economics somehow handling living things?

I think there is a very technical answer to this and a less-technical answer. The very technical answer is that in biology we don't have things such as

Noether's theorem, which tells us that in physics with every continuous symmetry group you have a conservation law. Weirdly, Emmy Noether coined that theorem in the 1930s, I think. It's the most beautiful theory in physics in my view. So, we don't have these symmetries or these homogeneities that in physics you often take for granted when you set out doing things.

So, you don't start with certainties?

Very, very rarely. Everybody would agree that the theory of evolution is a fact, and we would base that as one of the features or underlying principles of our work, but it's not necessary to involve the theory of evolution for all the stuff we're doing. And I think the way in which engineers, physicists, and chemists come up with certain models is fairly similar to our models — some of them are bio-physically motivated models, they are very similar, for example, to some of the models that physicists have used to describe the growth of surfaces. And there are some similarities which go beyond the superficial, but it's not always helpful to use this as a starting point.

We've covered some of this, but what would you say is the aim or the use of the model in your field — is it learning, exploration, optimisation?

It's two-fold: it's understanding and prediction. And the two, to different extents — you can't do both perfectly in a single context, you always have this trade-off. Certainly, in my field, there is always a trade-off.

Earlier on, you mentioned a simulation. What would be the relationship between simulations and the model? Or, can it be different relationships?

If we take the model as a representation of a natural process, then often the mathematical descriptions of these models are so unwieldy and complex that we can't make much progress with paper and pencil. We then have the choice of either simplifying the model to the extent that we can do some paper and pencil work, and get analytical solutions which will be right for all eternity, but for a model which might be too simplistic, we go towards simulating a more detailed and potentially better model, and then try to see how that agrees with Nature. I like doing both. There are also right now a large set of problems that we have not been able to solve with paper and pencil for a long time but that can be solved with computers quite easily, and in some rare instances it might become possible, based on the simulations, to come up with simpler models where we can get some analytical insights again for that. So, we can use the simulation to also guide the development of simpler models and try to see how much of that is captured.

So, it can be back and forth?

Yes — and it should be.

It is not necessarily just at the end of the modelling process, it's part of the modelling process?

Well yes, it's part of the modelling process, and even if you want your paper and pencil to do analytic work, it becomes much, much easier if you know what the true answer is. So, if simulation gives the true answer, then you know where you have to go. We think that there is a lot of scope to go back and forth between simulations and more classical mathematical modelling.

What has been the impact of the development of new tools and techniques on modelling in your field (e.g. such as telescope in cosmology, etc.)?

Without a doubt, computers. And our ability to use them better, build them better so that they are more adept for the problems we want to apply them to, and to analyse their output in a more systematic and statistically meaningful way. Just as we have to apply statistics to understand the real world, we need statistics to make sense of the simulations we make.

Moving now to questions of risk and uncertainty — how do you define uncertainty in your field?

Again, uncertainty is one of those words that does change its meaning the longer you deal with a certain element of uncertainty. For me, the most interesting aspect of uncertainty in my work is how it relates to the relevance that my model has for reality. So, there are some aspects of the model that I am uncertain about, or I'm uncertain how they relate to reality, and I want to get as much certainty about all the different parts of the model that we're looking at as possible. That is one type of uncertainty, and that is attractive to me. There are other types of uncertainty.

Would you describe this part of uncertainty — I'm being over-simplistic, but to be sure I understand — if you have a parametric model, will it be something similar to parametric uncertainty?

Yes, you're absolutely right. Parametric uncertainty has been looked at quite a lot, and that's something which I'm interested in. But what we're really interested in is **topological uncertainty** — so what different types of model structures could give rise to — that's the 10,000 versus the 40 million.

So, you have a possible class of models, and this is about the uncertainty within this class?

Yes, and that is the stuff that interests me a great deal. What aspects of the model I'm least certain about — that is something which interests me a great deal.

Coming back, you mentioned earlier that let's say you have this relationship between A and B and then B and A, and you said you have, say, 400 models telling you it is A influencing B and the rest, 9,000+, that it's B influencing A, so it seems more logical to believe in the second one. In financial applications, which is more my field, we have also these things, and it's also called model uncertainty, and people also tend to believe in the second one, but they will say maybe there is some relevant information that we cannot completely get rid of in the first one — even if it's just for a small part of the model. Because it might be just one model, but this is a very good model even if all the other models are against it. Do you know what I mean?

I do know. There is a point reached where you can fit differently. You go from classical statistics trying to fit your model to data to something called decision theory, where we try to see what other factors there are. For example, if we had this one model which was different from all the other models but would describe a potentially disastrous outcome for a patient, then we would try to see what is the cost of having the disastrous outcome with its small probability, versus the benefit that the other models would predict.

So, it would be something in some particular type of decision — a precautionary type principle would apply in that case?

I would hope so. It is not something which has occurred in practice for our work in biology. But I do think the precautionary principle is generally not a bad idea. So, certainly dismissing it out of hand is stupid. And then you just have to see...some people have said it's not a panacea, it can't be used to solve everything. But I think that in each case you would have to argue why you don't want to apply it.

You said you have different types of uncertainties, is this the first type?

That's the main uncertainty. Linked to this is the uncertainty, the most interesting thing is really all related to the uncertainty of the model, whether it's the assumptions that we put in, or the structure of the model that we have, and then there would be some uncertainty as to how our outputs are related to the outputs that the real world gives us. So, those types of uncertainty are the more interesting ones to me in my day-to-day life. Some of the work that we're doing has other levels of uncertainty in the outcomes, so, for example, we would not necessarily be able to predict the outcome of releasing these bacteria in the wild. So, that's something which would necessarily have to involve other people and — I don't like the word stakeholder as such — but people who have to have a say in that.

Would you qualify this as uncertainty or as risk?

There are risks and we are uncertain as to how likely these different risks are. Sometimes, there is a risk, but we are absolutely certain that it is not going to happen. And sometimes, we know that something is certain to happen, but we don't know how risky it is. In other cases, we know that something is certain to happen and we know it's risky. I think it's very hard to explain the difference.

So, would you compare this, let's take the example of the bacteria, as I think people can understand this quite well, would you say it's something comparable to climate change? You have a lot of risk and uncertainty.

I think there are certainly lessons to be learnt from the climate change issue. With climate change, I think that this risk is clear, and that the scientific evidence is also clear. But it's very hard to explain to people…it's very hard to punch holes into this argument because often the argument depends on, or you can represent the argument to depend on, many assumptions some of which people can claim are outlandish. So, how could you use tree ring data to re-construct past climate data, or ice-core data? And so it's very easy for somebody who is sufficiently persuasive as they are fraudulent to punch holes into these things. I think for us the risks are not as clear, and the uncertainties are not as clear. We can learn from past experiences how we should try to convey these uncertainties. But for me the case with climate change is you have got a couple of people who deny it, and you have got tens of thousands of scientists who don't deny it. It's very much like evolution — you find a couple of people who for nefarious reasons or plain ignorance deny it.

Going back to economics, the traditional difference between risk and uncertainty is about probabilities; with risk, you don't know the event itself but you would have an idea of the likelihood, the distribution; for uncertainty, you don't have a clue of the distribution.

Ah, ok so for us that's interesting. For us, the uncertainty would mean that we would only be able to make probabilistic statements. That is how we would view uncertainty.

We interviewed some chemical engineers and they had the same view.

The interesting thing is that often the precise shape of the distribution is not known, or the nature of the distribution might not be known, but what you can then do is you can apply a set of tools that even safeguards against uncertainty, about the shape or form of the distribution. Those go by the name of Bayesian non-parametric approaches. So, we would always say there is uncertainty associated with probability, but we don't necessarily know the shape of the distribution.

And the common belief would be, the more you learn, the better the estimation you get?

Yes, but we'll never reach certainty because there's always going to be residual uncertainty and an overwhelming richness of the potential models.

Because learning is infinite?

Yes, exactly. There will always be a high level of uncertainty for the foreseeable future.

That's interesting, because it seems that in your description of uncertainty, you have really two main sources: you have the uncertainty of the reality, and the model could help to understand this; but you have also the fact that you have all these classes of possible models, or good models, and then you have uncertainty among models.

Yes.

And risk?

Risk for me is something that has societal implications. So, it's very hard for me to dissociate that. We work purely theoretically in my view, but we work with experimentalists and clinicians, and some of what we're doing might have risks for a patient; some of the work, not so much we, but the work that others are engaged in, might have risks for large groups of patients.

So, while uncertainty is sort of a component by itself, risk is always specific to somebody who is...

Suffering. Yes, for me, there are risks attached to it and that means there are potential detrimental outcomes. So, I take in a decision theoretic framework, there's uncertainty as to how bad it will be, so there is a risk that for me it's risk versus benefit and uncertainty versus certainty, so to speak, and so that makes the distinction quite a lot different...my use or interpretation of the word is different from what you would encounter in economics, I can see that easily.

So, would you say that risk is somehow associated with the outcome of the model? Can it be used as a criterion to select models or to disregard models?

So, this has been done in decision theory, where you have to decide. You have all these different model predictions, and how would you then plan intervention in a real-world system? There was, for example, recently this mitochondrial donation question, which was big in the public — where the maternal egg cell has a defect in the mitochondria, the power-house of the cell, that gives rise to devastating disease, and the UK has now licensed the next step of research, and soon application, I believe, of using mitochondria from a donor egg cell and inserting that into the egg cell of the mother. People thought there were high risks associated with this, and then we tried to see what we should do given that there is a potential risk, and there was this very, very long drawn-out, several inquiries, the Wellcome Trust had a huge inquiry; and in the end the Wellcome Trust and lots of other people decided quite rightly that it's something they should back.

That's really interesting: it seems that uncertainty and the relationship between model and uncertainty, whether it's a class of model, is something somehow between just the researcher and the model; whereas the risk is something that has to be appreciated by an outside body?

Yes, that's what I meant before when I said it has a societal implication.

The researcher cannot really assess it himself/herself, it has to be assessed by somebody external?

Well, I have a view of certain risks, but there are times I can probably assign a certain probability for different types of risks, but there might then be other considerations which need to be taken into account, whether it's worth accepting this risk, I mean people have awful choices to make when they are diagnosed with certain types of disease — do they have to follow this treatment or that treatment — and doctors can give them outcomes saying 65% of people of your age and ethnicity, gender, will have this outcome, 34% will have that outcome, and 1% that outcome, so you want to go for this option. Well, I might have strong opinions about what the risk is, and I might even argue for one thing or the other, but it's not just me being affected, unless it's my personal risk of having a certain disease and so on.

Yes, that's very clear. The last question, which is a more personal question: what is the experience or result in your field which has had the most significant impact for you, and why? It might be something of your own research, or it might be something done by others, it could be anything.

It's quite hard to say because what I like about science is the typically incremental advance that we have, so there's always a little bit of advance, occasionally there's something which seems fantastically big and groundbreaking, earth-shattering, but it rarely ever seems to be… I mean there are lots of things I find fascinating which have occurred over the recent past. I do think that what has shaped my personal life and outlook on work is that people have been brave enough to consider that mathematical models can be applied to biological processes. And that started seriously in about the 1950s, 1930s potentially, with genetics, ecology, epidemiology, and then in the 1960s developmental biology followed suit and so on. So, it's a very unglamorous thing, but it's the sort of thing which somebody thought of. Nature is complex, but we can still use mathematics to tame some of that complexity and understand things better, whereas previously people had looked just at catalogues of biological diversity and they said there are these many animals of this type here, these many black rhinos, these many white rhinos, and they hadn't looked at how the number of rhinos depends on other factors and so on. So, that's quite a dull answer, in the sense that I find it hard to pick out something which is fantastically earth-shattering.

I think it fits really well. It's funny because the period you mentioned, I would say it would be the same for economics — all the mathematics applied.

Well, there's a lot of overlap. Morgenstadt and Neumann were interested in very similar things.

That's really interesting, to see the parallel, I didn't expect that.

Well, there is a lot of stuff that has been done; if you look at books on game theory, they give equal footing typically to John Maynard Keynes and Smith and John Nash. So, in terms of who came up with ESS, it's just a theory that has roots and applications in both biology and economic theory. Economic theory has shaped biology a lot, and I assume vice versa (as is currently the case, e.g. in the application of ecological theory to the analysis of financial networks and the stability of markets.

Now, I would say biology shapes economics because contagion and things like it have a massive impact.

Yes, absolutely. If you look at Darwin, one of the books that he had was of Lyell, and also Malthus's *On the Principle of Population*, that's also based on a then [sort of] economic theory and computation for limited resources and so on.

Chapter 13
A Conversation with Nigel Klein

Can you describe your field briefly?

I'm a doctor, and I'm one of a relatively small group of doctors whose job is split 50% doing clinical work within a hospital and 50% doing research within a university. A lot of people do predominantly one or the other, but I genuinely am 50% — I'm paid half by NHS and half by the University.

And your area of research, or practice?

I work with children, and my speciality is infectious diseases and immunology, and my research is in the same area — I research the conditions that my patients suffer from.

Present a paradigmatic example of a model in your field, describing it in terms that are accessible to non-experts.

Do you mean a model of the way I work?

Something that you're studying. Models that you use in your work.

So, like what I was just talking about — the mathematical model?

No, like for instance, if you're talking about children with HIV, if there is a model around this, or what you would refer to as a model?

Well, we have model systems for operating clinics, for example. I am very keen on establishing multidisciplinary teams to tackle complex diseases. So, for example, we look after children with HIV and developed a multidisciplinary service with clinical nurse specialists, psychologists, nutritionists, doctors as well as medical students; and the idea is that by having all these specialities together in one place, you can provide all the care that you need. We have used this model for other diseases, for example, I also look after children who have fevers. Within our clinics, we have the doctors who might be referred patients with fevers. This way, patients don't waste time having to see multiple doctors at separate appointments.

I think this is not quite, I think what I had more in mind, maybe it's not called a model, maybe protocol? So, if you have a patient, then from the diagnostic to treatment…

Ok, so we would probably call them guidelines, clinical guidelines. So yes, we write clinical guidelines. So if we just take children with HIV: we have developed protocols across the whole of Europe, of how to manage children with HIV. We then publish them.

So, for instance, if a new patient arrives, then you do some tests and there's a diagnostic, and then once the diagnostic is established, you have different "scenarios" and then you follow those 'scenarios' to have the treatment, or care?

Yes, we try to do everything in an evidence-based way. So, for example, the guidelines that we have will depend on the age of the patient — we treat babies differently from older children — we'll have a different set of guidelines, different things you might want to do, different tests you would do. But we establish them and then publish them and then we update them all the time, to take into account all sorts of new information.

So, would you say, is it like a decision tree? For example: is my patient a baby or older? If it's a baby and it reacts to the treatment, then we do this; if it doesn't react to the treatment, then we do something else…?

Depending on which disease it is, as I look after a number of diseases, some are much more like that. So, for example, there's a condition called Kawasaki disease; it's a rare condition, but we have quite a clear set of decision trees: if you come in with *x*, you must do this, and if you have *y*, you should do this. Sometimes it's not quite as didactic as that, sometimes it's guidance. So, for example, we say that all children who are diagnosed with HIV at less than 1 year should be on treatment. But if they're older, we may be more flexible. It just depends on whether they're unwell and what their immunity is like, and that will influence whether to start treatment or wait until some point in the future. If we see enough of one disease, we can develop guidelines and then it's available for other people to use and treat in the same way. Now we have done further work and indeed we now recommend all ages are treated with medication.

So, is this guidance, or guideline, based on your experience and also the experience of other teams?

Definitely, yes. It's a bit of a mixture. In the areas that I work in, I will contribute something, but other people will as well. We try collectively to do the best we can for our patients.

We'll come back to this, it's really interesting. So, if say a patient comes with symptoms you have never seen before, how would you start?

There are basic principles that I will have learnt over the years. So even if, as occurs most weeks, I see something that I've never seen exactly the same before, I have an approach that can be used. For example, many people who are unwell have a fever, but it may not be obvious why. I would start by thinking about which conditions may give rise to fevers. I would then look for other features — this may include a rash for example, or you'd look to see if the patient has pain anywhere. All the time one is compiling bits of information, so even if you haven't seen something exactly the same before, you can integrate the information and try and identify what sort of problem it is. For example, you may identify that the problem is in the heart, or in the brain or a problem in the immune system — you can always find ways of approaching a patient systematically, to identify what the problem is.

That's very interesting. So, I will not use the word model, because it does not quite work…I'll use guidelines?

Guidelines or protocols. I like guidelines, because it's telling you they're there to guide you. The problem with protocols is that protocols tell you what you must do, and for me it prevents people from using their brain.

I really like guidelines. We'll come back to this later when we come on to uncertainty, but I really like that guidelines leaves room for the rest.

Exactly, I completely agree. So, that's my view. You say, if you really have no idea of what to do, then you can take these very literally, but if you're experienced, you have more flexibility, I can say you prefer to use this drug rather than that drug, providing the principles of the guidelines are maintained.

So, with this in mind, could you explain why a guideline is needed?

There are three reasons: the first is, it's an education for me; because to develop guidelines you have to understand what you're dealing with, and you've also had to read all the necessary information. So, I think it's a very good personal education. Secondly, to develop effective guidelines you have to go one step further with the information you have obtained, to use the information for a practical benefit. So, it's not just a set of ideas, you actually have to put the information in a form that allows others to proceed if they see a patient with the condition. And I think if you've done that, and you've done the job well, then other people can then use those guidelines even if they haven't got any experience in the area. So that's what you hope, that because you've done a good job, then the less experienced, or even completely inexperienced individuals, will be able to look at the guidelines and follow them. In time, they may alter their practice; but at least they have got something to start with.

So, this is really a bridge from observation of a patient onto the treatment?

Yes, the guidelines will help someone to make the diagnosis and then provide the appropriate treatment.

Does mathematics (or quantitative methods) play any role in the guidelines?

Definitely. It has a role in some guidelines. A lot of what we do is based on clinical trials. For example, I mentioned Kawasaki's disease; it was discovered a number of years ago that if you gave something called immunoglobulin, if you gave it intravenously (so it's called intravenous immunoglobulin) and if you gave this within a certain time of making the diagnosis, then one of the complications, damage to the heart, can be prevented. This data came from a trial: with this medicine, patients did much better. To do a trial, you require quantitative methods, and we have very good statisticians who are involved with these types of studies. But mathematical modellers also have an important role. In our treatment of children with HIV, one of the things that I was very keen on doing was seeing whether I could develop a way of making a prediction about what their immunity would look like when they became adults. So, we got some very clever mathematicians together and they developed a model which is now published, and forms part of our guidelines. We have a graph in the guidelines where you can look at someone's age and one of their immune tests, and with those two bits of information it is possible to make a prediction of what their immunity will be when they're 20.

So, you would say the quantitative methods — the mathematics, statistics — are ways of narrowing the guidelines so that each one is more appropriate for the patient?

Yes, and also to help provide the evidence needed for the guidelines. People are very keen these days to assess the quality of evidence used for guidelines. There are now grading systems in place for judging the quality of the evidence. There are some diseases that are so rare that the quality of the evidence is poor and we rely on the opinions of experts. That's considered low-level evidence. The best evidence comes from large clinical trials of lots of patients so that you see big differences if the treatments work. So, we use all the evidence that we can. But there's no doubt that our mathematical approaches are getting better and better. We try to use them more and more; even when we don't have very many patients, we can use mathematical approaches that are specifically designed for trials with only small numbers of patients.

Beside these mathematical formalisms, or tools, what constituents are part of the guidelines? You mentioned when there is an extremely rare

situation, you have to rely on the impression, or the expertise of the doctor, but what else?

It would be clinical trial data; there may be basic science data; you may say although there's no evidence in patients as yet, in animal experiments this drug was found to do this. So, in this case, while we can't give good evidence for use in humans, we can say that it's something you could consider based on the work that we've done either in the test tube or in animal experiments.

What is the role of language in guidelines?

Language is key. If you don't communicate well, you can't get your message across. The clearer your guidance, the easier it is for people to follow. I noticed this two days ago when I had a question from someone in Portugal who's read our guidelines about one of the conditions, and wanted some clarification. In the future, we will have to make this clearer, but they also said that apart from that the rest of the guidelines were extremely clear. These days, in part because of the Internet, guidance may be read by people who are not necessarily English. Even though English is the main international language, we have to try and make our guidance as clear as possible, and try to generate diagrams and tables that are clear, because we want to get our message across.

You said earlier, you prefer guidelines to protocols, because protocols are a bit 'you must do this,' whereas guidelines are just guidelines; so in the language I guess you don't use terms such as 'you must,' you give a bit of freedom of interpretation to the person reading the guidelines. Do you control the space for interpretation?

Yes. I don't like de-skilling readers. It is important to allow individuals to use their own experiences. For example, coming back to Kawasaki disease: in Kawasaki disease we know that the earlier you treat, the better the outcome. So you want to phrase your guidance to indicate that it is better to give treatment than not to give treatment if you're not sure. Whereas for some of the other aspects of care that are probably less critical, less dogmatic statements can be used, such as 'if the patient is not getting better, then you can consider the following…,' whereas for critical issues such as making a diagnosis, you might say, 'making a diagnosis can be difficult, if in doubt, start the treatment, because if you don't and it turns out they have this condition, you may have missed that window of therapeutic opportunity.'

Do you have any qualitative aspect in the guidelines?

Yes, I think in many conditions we can't be quantitative because we don't have the data. But we also work in areas that are only qualitative as they cannot be easily quantified. Examples include whether a patient is feeling well or whether the family feel they've made a full recovery. For these less quantifiable aspects, qualitative tools are being developed to try to make an assessment more robust.

From what you said earlier, 'if the patient is still unwell, you may consider…' — still unwell, for instance in the case of a fever, you might have a fever going down but not becoming normal — that kind of thing?

Yes, that kind of thing.

Back to models: models are often said to represent a target system. Does this characterisation describe what happens in your field? So for instance, you have climate models representing the climate of the Earth; for guidelines — can this be rephrased in terms of guidelines? Do they represent the state of evolution of the state of the patient, when a treatment is applied?

I suppose you *could* say that the model aspect of guidelines could be that it's two-way: someone has a problem, you've written guidelines to provide an approach to this problem, you initiate a course of action, and then you get feedback to see whether or not its working; and if it isn't working, then you go back to your guidelines to see what is the next thing you should consider.

That's really interesting, because in most other fields you have the model, and then somehow the decision of the decision-maker comes after. The model is for the decision-maker to take the decision. Here, it

seems that the guidelines do not only understand the situation, but they also incorporate the decision at the same time.

Yes, it does. So, feedback is critical. In medicine, all the time you're gauging whether or not things are moving in the right direction.

And so this is a massive block between the observation and the end-result...

I've never thought of it quite like that, but that is what's going on. You're assessing lots of different types of information all the time. But what you do in your guidelines is focus on key bits of information; so in Kawasaki disease, fever is critical — in fact, fever is part of the definition; so, if you're not bringing the fever down, then we think that's telling you something about not being on top of the disease; you then will try something different. So, we are always assessing responses to our actions. Indeed we are now testing new approaches in a European Trial in this disease to make sure we have the best treatments.

So, the role of the doctor, or medical consultant, is key, because he's there from the beginning to the end...?

Yes, absolutely. It's a continuous process.

What is the relationship between model and theory? Do you have an 'ideal' patient, or even if I may use this term — do you have a theoretical patient?

You may. I think it depends on conditions. Sometimes, you may not have any...there's no good information on what you should do. I'll give you an area where this is the case: we increasingly have what are called 'biologics.' Biologics are a new class of medicine. They're usually very expensive. They've often been developed based on a theory about the way the body works, because they're very, very specific. In the past, we used many drugs that had a broad action, such as steroids. Steroids have a big effect on many systems within the body; they're very effective, but also quite dangerous, particularly if you take them long term. So, these biologics are meant to be much more specific. They will target a particular molecule. The company that produced the new drug may not have known definitively that this molecule will work in a particular condition, but theoretically they think it could do, based on all of the science that's known about this molecule. So, you have a theoretical approach initially; you then start using it, and institute various ways of assessing whether or not it's working or going to work. So if, for example, you produce some guidelines to manage a particular disease, a new drug is produced; you haven't yet had a chance to test the new drug, but theoretically, you think it could work very well. You can then either try it out, or try and formulate some new clinical trials to assess its use in this disease. But the reason you're using or developing a new drug could be on theoretical grounds alone.

So, would you say that theory in your field, if we look at things time-wise, is somehow prior to the guidelines, it is part of the clinical trial or things that come before establishing the guidelines?

It could be at any stage. You're updating, you're re-thinking and planning for the future. Feedback is really key. Homeostasis is a term used to describe the processes that act to maintain stability. In biological systems, there are always some fluctuations in and around the homeostatic set points. One could argue that perhaps what we're trying to do, as physicians, is to try to bring people back to their homeostatic point. But to do that, we're observing things to help us to intervene appropriately, to bring you back to that point.

I guess this is quite clear, but what would you say is the aim or use of the guidelines?

The aim ultimately is to improve the treatment of the patient, that's the ultimate aim. Either because the process of developing the guideline has then made you think more about how to do this better, or purely as guidance for someone reading it to help them do it better.

Is there any use of computer simulations?

Yes. Many models use computer simulations. It's amazing really. I'm not a mathematician, I just work with them. What's amazing is watching the way the simulations work. So, for example, you can put information in to these models, and the simulations will then construct the outcomes for you, and you can see how that works.

As how a patient will react to a certain treatment?

Yes, some of our models are becoming very personalised — it's called personalised medicine, it's the way things are moving. So, what you may be able to do is to take some key bits of information and put them into your model and then the model will simulate what will happen. There are other simulations that you can use. For example, every year we have to undergo training in managing cardiac arrests. So, if someone had a heart attack, we have to know how we would deal with it, even if we're no longer ourselves dealing with that aspect of medicine. So, they now have simulation software, it's a computer system — the computer basically drives the scenario: suddenly, it says the pulse has stopped, do something; and you start doing it and it's giving the feedback, such as the person's dying. So it's a different type of computer simulation, but we are using these simulations more and more.

And are the computer simulations useful in order to give precise guidelines?

Yes, to aim the guidelines to an individual, or to improve the guidelines. If we can use computer models, then we will.

What is a good model? What is a good guideline for you?

I think a good guideline for me is one that uses available evidence but also encourages further research/trials as well as stimulating individuals to do things better. If it's too didactic, it stops people from thinking. So for me, a good guideline updates you as to the best practice, but also encourages people to think creatively about how you could do things better.

That's the first part on models. Now we move to the second part, which is about risk and uncertainty. We touched on this with the re-naming of guidelines. How would you define uncertainty in your field?

Uncertainty is at every level, from the point of making a diagnosis, to what is the best form of treatment. From a young age — I remember my first night on call in hospital, and I remember everything was uncertain; things that now I wouldn't think twice about, such as the very first time I had to write up a paracetamol, not just go to the chemist and get it, but to write it on a prescription sheet, making sure the dose was correct. So, I've been trained to live with uncertainty.

Uncertainty because you're dealing with humans?

Yes, you're dealing with a biological system, and so there's uncertainty. One realises with time that you live with it and accept it. It's part of my job, the uncertainty, and what you're trying to do is to put as much certainty into your practice as you can. So, I'm quite comfortable with uncertainty, but I recognise that that's because I'm trained to feel like that. The uncertainty is a driver for both my clinical work and my research.

And the way to reduce the uncertainty is through experience, and also through the guidelines? How does it work?

Yes, what you're trying to do is to initiate research or acquire information to reduce the levels of uncertainty. That's a major driver for my research. Generally speaking, I do research in areas that we don't have answers to — I try to answer questions that would reduce the uncertainty. However, other people may already have the answers. But if no one else has the answers and it's in my area of expertise, then I will write my next grant to try and answer these questions, to reduce the uncertainty in the future.

And how do you define risk?

Risk is a very interesting question. In the medical field, the risk is that you do more harm than good, you're always thinking about that. Sometimes that's easy — you know that if you don't do something, the risk of *not* doing something is so great that you might as well do something.

Like you mentioned in the Kawasaki disease.

In Kawasaki disease, yes, the risk of giving immunoglobulin is relatively low; there are some side-effects, but actually there are very few serious side-effects. If you don't give it, then the risks can be high. So yes, the risk-benefit of giving immunoglobulin favours giving it. But if, for example, a new drug is produced and someone says this should work better than immunoglobulin, but we haven't done a trial yet, it would then be a risk using that drug. So, all the time you're weighing up the risks and benefits. It's one of the areas that I think drives a lot of decisions in medicine. You may or may not know that people are very worried about antibiotic resistance. An important reason for why we have antibiotic resistance is that we're not using antibiotics very well. For most people, the risks of giving an antibiotic is small — you can get rashes, you can get bad reactions to the medicine, but with antibiotics, most of the time it doesn't cause any obvious side-effects. Yet, by giving those antibiotics you will be selecting out bacteria that are resistant to those antibiotics; it may not affect the patient, but the patient then carries with them resistant bacteria and they can spread them to other people. So people use antibiotics all the time because they *perceive* that there's very little risk.

So when I go around the hospital, I spend a lot of my time trying to persuade people not to give antibiotics. But it's hard, because people don't think there's much risk. There's a global risk as opposed to a personal risk.

A question — coming back to the guidelines and also uncertainty — let's say your team or you establish guidelines for a particular disease, and another team somewhere else in the world establishes another type of guidelines for the same disease, how does this work? Is it possible to have two different sets of guidelines for the same disease?

Oh yes, definitely. That's an excellent question. Yes, there are lots of guidelines. So for example, we have European guidelines for treating children with HIV, but the World Health Organisation also has guidelines; its guidelines are slightly more orientated towards Africa, whereas ours are more orientated towards Europe. But there will also be American guidelines, and there will be differences between theirs and ours. Uncertainty means that you can have more than one way of approaching a problem; and often there's a lack of evidence to define one particular guideline.

So yes, you will see multiple guidelines. Usually the basic principles are similar, but it's the details that are different.

And the details can vary because of the availability of some drugs, or…

Yes, availability of drugs, personal preferences, interpretation of studies — a study was done where if you gave a particular medicine to babies, they did better than with another medicine; but we know that this medicine tastes horrible, and so getting young children, babies, to take this medicine is very difficult. The reason why in the study it worked is that they employed lots of people and with all that support they managed to get the patients to take the medicine. But we know, when you don't have that support, its difficult to persuade patients to take the medicine. The people who did the study were very keen to use this medicine; but we know in practice it won't work because this study only showed it was superior because of the support. So, there are practical reasons why guidelines differ. Sometimes because it's in a different part of the world; sometimes it's because of the availability of different medicines — for example, in a hot country, liquids may not last as well as tablets.

So, you have this additional uncertainty in terms of the coexistence of multiple guidelines, and so the choice depends very much on where the consultant is, or the patient is?

Yes.

Can you, as a British medical consultant based in the UK, use American guidelines?

Yes, providing you're doing something that's based on some sound opinion, then it's fine. If the guideline was from someone who no one knows about and/or they don't have any evidence, then we should be suspicious of these guidelines. In some parts of South Africa, 85% of people go to traditional healers. If people believe in their traditional healer, then they'll follow that guidance even without any evidence. If American guidelines differ from European ones, but a Consultant has good reason to follow the American ones, they can do that.

Would you say that, if you have two patients, is it possible that you would use one guideline for one patient and another guideline for the other, or you would have a sort of systematic approach?

Personally, I would like to feel I have a systematic approach.

Last question, which is a personal question: what is the experience or result in your field that has had the most significant impact for you and why? It need not necessarily be one of your own results — it might be, but it could be more general as well.

That's a very tough one. I sometimes lecture in Hong Kong, and I had to give a talk on the 10 most influential papers that have altered my practice, from anywhere — not just my own work. And I didn't have any problem finding 10 very good papers. Can I use two examples? One of them I like because it challenges traditional thinking about something. It was a study that was done a few years ago in Africa, and based on work that we and others had done in Europe and the United States. We thought it would always be good, when a patient has a serious infection, to give them more fluids. So a trial was done in Africa with a group where they gave more fluids more quickly in one group compared to another group. And to everyone's surprise the group that were given more fluids did worse than the group given less fluids. This is still causing controversy; some people said you must have got the results the wrong way around, because people just didn't expect it; other people said they don't understand it, so they don't feel you can use this to influence guidance in Africa. But the reason I like the study is that it's challenging people to think more about simple things that they do — so giving fluids, people don't think that much about it, but actually it is really important and we have to understand how to give fluids. So, it's stimulated lots of discussion and it's stimulating new research. I didn't do the study, it's just that I enjoyed the study, it was a huge study, beautifully presented and executed, and my view is that it's been very, very important.

I think there are a few, in my own personal practice, perhaps the work that I feel most proud of doing is our mathematical work in HIV, because I think we've made people think differently about how to treat children with HIV. So, it's not one particular thing that we've done, it's more that it's taken people from a very rigid practice to saying, actually, if you don't think differently about it, you're going to miss opportunities. So, perhaps it says something about me, that I like work that changes the way people think. Particularly my trainees, I like to get them thinking. A friend of mine always used to joke that he always used to encourage the brightest medical students to become paediatricians, but as he gets older, he's now encouraging the best medical students to become geriatricians because then they'll look after him better when he is old. So, to some extent, I think if we train doctors well and make them think, they will improve medical care. There may be other things that I've done that have actually made a bigger difference, but these two examples challenged the way people think and, therefore one hopes, will ultimately improve the care of patients.

Chapter 14
A Conversation with James Keirstead

Could you describe your field briefly?

I study urban energy systems, which is basically the demand for and use of energy and its supply within urban areas.

From an engineering point of view?

Yes. My background is both engineering and policy — and I look at the intersection of the two.

Present a paradigmatic example of a model in your field, describing it in terms that are accessible to non-experts.

In any energy system, you usually think about the problem either from the supply side — what are the energy sources being used, how are they being converted? — or from the demand side — why do people want to use energy in the first place? And so you can think of example models for both of those categories. On the supply side, a good model would be an optimisation model, where you give it known demands, and known technologies, and the question is what combination of technologies and fuels would meet the energy requirements of that urban area at minimum cost, subject to constraints on carbon emissions, or other sorts of things like that.

And typical sources of energy could be like electricity…

Yes, electricity, heat, … In those sorts of models, we tend to put in the end service demands — so that could be electricity, space heating, hot water requirements, cooling — and then the sorts of fuels that would be coming in might be electricity from the grid, natural gas, maybe biomass, oil, things like that.

And so the idea is to minimize the cost function, with given prices for buying electricity…

That's right. And conversion technologies. So, as a simple example, if you have a home-owner and they were needing a space heating service, they could choose to, say, buy in electricity and run a heat pump, or they could buy in wood and use a wood stove, or they could buy natural gas and have a gas boiler. So, the cost function would include both the fuel requirements and also the choice of technology — the boiler, the wood stove.

With the help of this particular example, could you explain why a model is needed?

The main problem is that the possible conversion pathways quickly get too complicated to assess. So, if you have a large urban area, you have a combination of demands from domestic customers, from industrial customers; you would also have utility scale things — so, someone might provide a heat service, for example, that can go through pipes and service a combination of domestic and industrial customers, and if you include more than two or three options, it just gets too complicated to figure out which is the best combination.

So, the model helps you with that.

Exactly, yes.

And could you describe more precisely what a model is? (In particular, what are the different stages in the modelling process?)

I like to use Rosen's definition of a model, which roughly states that a model is a formal system that replicates at least some aspects of what we see in the natural world. Here's a diagram that explains the process.

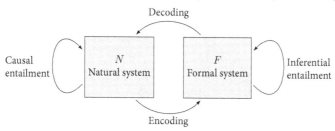

The modeller starts by observing the natural system and tries to figure out what its internal logic is; in other words, when I see x happen in the real world, does it lead to event y? We can then encode that observation in some sort of formal system: maybe a set of mathematical relationships or computer codes, or maybe just a mental model of what's happening. That simplified formalisation has some sort of internal logic that allows us to try out different what-if scenarios. The results of these analyses can then be 'decoded' and interpreted within the natural system we're trying to represent. Our formalisation might be said to be a good model if the process of encoding our observations into a formal system, experimenting with that formal system, and then decoding back to the natural system, accurately represents what we observe in Nature.

In your field, what is the role of mathematics, or quantitative tools, in modelling?

The models are specified as what's called a mathematical programme, linear programme, model. So, it's written as equations, it's formalized as a set of equations.

So, you minimize under a constraint?

Exactly. In practice, the way that is implemented is that we don't solve those equations analytically (using Lagrangian techniques or anything), but we use software — so we code those equations into a mathematical programming language and we ask a computer to solve the problem for us.

Does it mean that the model is that set of equations?

Only partly. The formal system at the centre of the modelling process is indeed a set of equations, expressed in computer code. But the model as a whole contains the encoding and decoding processes, the translation between the real world and our stylised equations.

Besides the mathematical tools, do you have any particular constituents as part of the model? Any particularities?

Do you mean stakeholders as constituents, or do you mean component parts?

Component parts of the model — parts of the model beyond the equations.

Yes, the data would be the biggest one. We developed a model platform — in other words, a series of computer models, datasets, user interfaces, etc., for a single problem — which includes more models than what we're talking about now, but a common theme is that there was a data library. So when we would run, for example, this optimisation model, it would pull the relevant capital costs, performance data, and whatever out from this data library.

The "model," in an epistemological sense, is not just the optimisation model, which is a set of equations and the data that informs the calculations. The confusion is that the word "model" is colloquially used just to refer to the formal system part of the Rosen diagram. The modeller will be doing the encoding/decoding process, but they may not be aware that they are doing it, and that that is part of the overall model.

And this data library is something that is built upon historical data, or statistical data? Is it raw data or worked data?

There would be some working, but not a lot. It would basically be looking at manufacturers' catalogues or historical statistics, and then really the transformations required would just be those to match the parameterisation in the model. So, you may get a statistic that's measured in energy consumption in barrels of oil, for example, but we need to know it in kilojoules, so we make that conversion.

So, to be sure I understand: you have the mathematical equations of the model, then you calibrate your model on this data — so you use this data as input in your model and then...

Well, calibration is not quite the right term...we do a calibration step, so maybe it's worth clarifying; the model has lots of parameters that go into it and it will give you an end result; some of those are these numerical values that come out of this database, some of the parameters will be things that are not obvious, like — the cost of a gas boiler is obvious, you go look it up and you put that number in — but there will be things that will be closer to a calibration constant where you need to pick the values for which the number the model gives you is about right. So those sorts of things we do by trial and error, by looking at known cases, testing the model against those cases, and then adjusting that parameter — some sort of fitting process, but it's not a fit like a statistical model.

And is there any role for language in the modelling in your field?

As in natural language analysis, or just in the way in which we describe problems?

For instance, if say, you have a new problem to consider, is the formalism in terms of words an important part of the modelling...or not really?

Not so much. We do have a formalism which is really underpinning the data model — so we have an ontology representing the different types of objects in the model system, the properties that they have and the way they interact with the parameters in the equations. In other words, the data model is essentially an outcome of the encoding process; it's a design decision about what types of objects will be represented within the formal system and how we will represent their attributes. So, if we have a new problem, the question is...for example, we're looking at City X now, how do we take the data we have from that city and package it up, if you like, so that the geographical units of study match with those that the model uses for its calculation? So that's probably the closest thing to a common language.

Is there any role in your field for qualitative analysis? I'm thinking that some of the things you're modelling depend on, for instance, householders agreeing to take up whatever measures — Does qualitative analysis play a part at all or not?

It does…my background is in civil engineering, and then my Master's and PhD are in energy policy, and those theses both combined interview questionnaires and statistical modelling, like standard regression, modelling and choice modelling, etc. And we've also done a couple of papers about urban energy policy, like comparative governance-type stuff, and that was all interview-based, trying to understand, as you say, why people do the things they do; and some of those decisions, like whether a household will choose to adopt technology X or Y, that can be modelled fairly easily as a statistical technique, as a quantitative problem; but other things, like how the mayor decides this issue is going to be a priority next week, is something that you need to read about, talk to stakeholders, see what their opinion is; no one really knows what's going on, but you can get a rough idea.

So, some of that can feed into how you develop the model, or it might affect how you perhaps describe outputs, or the caveats…

Yes, something like the qualitative work is quite useful for setting out what the scenarios should be. So, if you know that there's been an undercurrent of debate about should we go down this road or that road, well then, you set up two scenarios that follow those and you do the quantitative analysis to investigate what those would look like.

And the common language would be somehow borrowed from engineering, and energy,…

Yes, it's a bit made up, in the sense that there's no well-accepted standard, but yes, certainly if you told an expert in this field and said we're modelling this city by saying it has a number of discrete zones, there are processes, there are fuels, …they'd say, 'yes, I understand what you're talking about.'

We will discuss this further later when talking about risk and uncertainty. But, would you consider scenario analysis as a qualitative tool in the modelling process?

Yes, I think I would. A scenario may be expressed quantitatively, i.e. these parameter values, these calculation options, but you will often find that scenario analyses are introduced with a qualitative narrative about why these formal choices were made, why they represent a coherent view of the world, and so on.

How important is notation?

I suppose a quick answer is very. One needs to be clear about, for example, what a value of x = 10 means in the real world. But on the other hand, I've not seen consistent notation between the papers produced by different research groups working on urban energy systems.

Models are often said to represent a target system — for instance, climate models representing the climate, or typically a selected part of the world. Does this apply to your field? Is the model to represent a target system?

No, not targeted at a single system. The model design is more flexible and aimed at a class of target systems (cities). So, we intentionally designed the tool so that it is generic — that any urban environment, and actually even if you abstract further, any spatial system, could be modelled with this sort of tool. In practice, when we do case studies, we need to situate it in a location, so we then modify some of the data inputs, to represent that target system. But the only changes we would make to the model itself would be that we might add some new equations, new constraints, that are only applicable for that one target city.

And for you, what would be the difference between theory and the model?

Well, …a model would be a formal representation of the theory. So you could have a theory about the way in which, for example, you think heat flows through a network or something like that. But the model would be the equations that represent that theory and provide you with some sort of calculation or testable hypothesis about the real system.

So, the theory is more abstract?

Yes.

And you might have different models for the same theory?

Yes, absolutely. In practice we don't, but in theory, it's possible.

What would be the aim, or use, of the model — in terms of exploitation, or learning?

In the case of our urban energy systems software, it's quite a normative tool. It's really saying this is the way a system should be designed to meet such and such a goal. We know from other cases the results of that sort of normative modelling are very rarely if ever seen in the real world, because things don't get decided on least-cost optimisation, but it's still a useful framework.

It's an aim.

Yes, it's an aim to inform stakeholders, policymakers, technical consultants, on what *could* be done if you were able to be a master planner and rearrange everything just so. In some of the studies, we've said it's really almost a benchmarking tool — that you can judge the ambition of the design against this optimisation…

And you want to minimise the discrepancy?

For example, yes.

If you use computer simulation, which you mentioned earlier, what would be the relationship between the model and the simulations?

The model is the formalisation of that theory, and then the computer model, the precise code we write to express that generic model if you like, would be a formalisation of the formalisation. So, you could formalise your view of the system with just the equations, and then the computer code would be one implementation of that. And you could have multiple implementations.

So, would you say that the computer simulations help you to better understand your model? Or somehow you have the model first, the computer simulation, but then can you go back to the model after that?

Yes. There's an interesting question about what *should* a person do and what *does* a person do. Yes, certainly in practice, we would have our computational tool, we would run it, look at the results, and say, ah that doesn't quite look right, and then we re-think the computational tool. A bad habit that a lot of people have in this field, and you may discover this over the course of the interviews, is if your tool is computer programming, there's a tendency to want to get started on the computer programming right away. So you don't necessarily deliberate as long on the theory, and then again on the formalism, before you start the third part of actually implementing.

So, the implementing becomes part of the modelling — almost immediately?

That's right. So you start building and you do little prototypes, and you think that doesn't look quite right, so in effect you're doing lots of these small little feedback loops with your understanding of the problem.

So, it's a bit like people doing a bit of an experiment in a lab and then going back to the theory.

Yes, that's right.

What is a good model?

A good model is one, I think, that people understand — and by people I mean people running the model and the people looking at the results of the model. And that it delivers some sort of useful knowledge. It could be as simple as needed or very complicated — as long as people understand what it's doing, and the value of that result is there…

And the limitations.

Yes, exactly, that includes the limitations. And that is a good model.

Coming back to some earlier remarks you made, what has been the impact of the development of new technologies or tools? (e.g. in cosmology, telescopes, computer simulations for DNA…)

Primarily to extend the complexity of the formal system. If we can compute with larger datasets, more quickly, with new programming languages that allow us to elegantly express complex ideas, then there is the potential to build a formal system that represents more and more aspects of the natural system.

So, now moving on to a new set of questions about risk and uncertainty… How would you define uncertainty?

Uncertainty we divide, I suppose, into epistemic uncertainty — whether you actually understood what the problem was — and parametric uncertainty — so if one of the parameters in the model is what's the cost of a gas boiler and you get that wrong by £100, does that change the answer? Those would be the two kinds of uncertainties that we would consider.

With the first one, if you don't understand the problem, what does that mean exactly? Would it be, for example, you thought it was a domestic demand for fuel and in fact it was an industrial demand, or not that kind of thing?

Yes, so if you thought, for example, that you could model the problem as an optimisation problem, but you realise that actually the answer to this question is determined by a complex system approach that is not a top-down optimisation, it's the product of thousands of people making individual decisions. So, that would be the distinction.

So, you see this uncertainty as 'shocks' to the model rather than part of the model?

Yes, ...we don't do this ourselves, but we know other people in the field that do this: who would include the parametric uncertainty as an integral part of the model formulation. So, they would have a stochastic formulation-based model that would explicitly try and find an optimal solution under uncertainty. Whereas we would base a deterministic tool, but do Monte Carlo with it, and use that to explore the uncertainty.

And so you would test somehow the robustness of your solution when you have evaluation in the parameter?

Yes, exactly. I've got some images from papers I can show you.

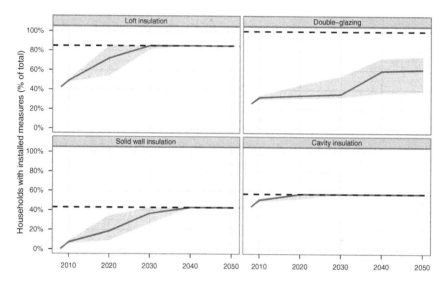

So how would you define risk?

A risk, I think, would be on the model interpretation side.

That's really interesting. So the uncertainty would be more part of the modelling, whereas the risk would be in the taker of the result and the risk in interpreting the results?

Yes. I suppose there would be another commensurate risk to the modeller if in making the formalism you forgot something or you were ignorant and misplaced something and so when you present your result to someone, you've missed something big, which is then a risk for you as the modeller, versus a risk for the decision-maker who's using the model results.

So, how can you reduce this risk, especially for the one using the model?

There's the 'in theory' and the 'in practice' bit! In theory, you develop the model in conversation with the users so that they know what's going on; in practice, that's not always possible. So what you do is…because we know people, even though you tell them the caveats, they don't always fully take them on board, so what we tend to do is really filter down what we tell people; so, if we ran months' worth of simulations and all the rest, we'd probably distil it down to two or three points that we know were pretty much bullet-proof, to try and…just to not try and confuse the picture by bringing in too many of the caveats, like 'sort of,' 'well this, maybe, but, kind of,'…

We spoke about robustness earlier. Of course, you have robustness towards an error, a misapproximation of a parameter, but let's say for instance, you have your model for given data, and then suddenly the price of oil goes mad — do you have ways of checking the robustness of your conclusions towards big shocks in the input data, or is it something you don't really look at?

We try and capture that either in the Monte Carlo or explicit scenario analysis. It just depends on how big a set of possible futures you're looking at.

So, you can do scenario analysis in this worst case scenario, or positive economy, or…?

Yes, exactly.

And so then you have different scenarios and this is also a representation of the uncertainty in a way…

Yes, that's right. With a small number of discrete scenarios you might try to map out the extremes of the uncertainty. You often see reports with little 2x2 diagrams, mapping out four scenarios that try to cover the widest possible range of outcomes. But another interpretation would be to say that a scenario includes even the smallest change in a numeric parameter's value. In that case, you are effectively saying that the uncertainty is limited, that the use of scenarios is just to explore the variability of the conclusion in a small area, rather than questioning the bigger picture.

Your models are mostly for policymakers and decision-makers. So, is it a challenge to present your results so that they are still scientific, but understandable by them?

I think so. I think the biggest challenge with this sort of method is really like a very fundamental thing with the method: it says the way the world should be if your goal is to minimise the cost, a cost which combines costs paid by lots of different stakeholders. So, it's a situation that no one ever really faces, and so you're already so far off something which could be called scientific that a person doesn't really lose a lot of sleep over it. You know that you've done the model correctly, you've tested that the mathematics works the way you expect it to, you've done your best to make sure that the parameter values going in were correct or at least good approximations, but then you're still left with this fundamental problem of is this actually a good representation of the way the world works and the types of decisions these people are facing?

We try and phrase this kind of modelling within the notion of post-normal science, trying to differentiate between really narrow scientific uncertainty versus professional consultancy, and other types. We do look at statistics a little bit — the difference between uncertainty and sensitivity analysis.

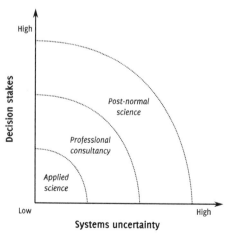

This is something I would like to understand better: your parameter uncertainty I think is somehow related to this idea of pushing the parameter...

Yes, and anyway, we show them how to do it and how it's applied.

So this is — when you say 'what is a model,' I always use the diagram on p. 269.

Rosen is a mathematical ecologist, and he says you have a natural system, and that has some internal logic, it means the world works the way it works, so you don't know what it is, but you know it's there somewhere. A model is IF you observe the natural system and its own internal logic and you can reproduce that by encoding that natural system in your formalisation, that has its own logic based on the axioms of mathematics or whatever, and then you decode the result; and if you basically get the same by going through this outer loop, as you would just getting the natural system, then that is a good model of the natural system.

So, when we were talking about risk, I was mentally thinking about these two things. So, there's a risk to the modeller that you encode it incorrectly, and there's a risk to the user if you don't decode it and understand the limitations of the formality.

So, that's what I use for 'What's a model?' And this, as I say, is for the post-normal science stuff. To the students, we explain that if you're doing, say, projectile motion here in a lab, well, you know basically the mass of the projectile, you know the angle and if you make it a bit longer, the consequences aren't that big. So, that's applying the science. If you were a professional hunter trying to hit something with a bow and arrow, that's 'consultancy,' because there are uncertainties that you need to experience to guide you to know how to use the model. And post-normal science in this paper, they say that's stuff like climate change — it's got such high decision stakes and is so uncertain, that for any numerate analyst to come in and say this model says the answer is 12 is just completely ridiculous, and so you should be bringing in other types of knowledge to better understand, such as the diagram on page 279.

And then we get into what is uncertainty analysis for us, if you have uncertain parametric inputs, you draw them from a bunch of distributions and you see what the model's output looks like. So, a simple example, where you pick some variables, you give them some sort of uncertainty, and then you go through and calculate what the result of the model would look like.

It really depends on the technique then…because for the bottom one there is almost no uncertainty, right?

That's right. And this — these are the results from thousands of runs of this optimisation model; and each run is basically coming up with a pathway to 2050, choosing to install different technologies at different times, and the nice policy takeaway from this is that by 2030, for cavity insulation and loft insulation in almost all of the scenarios, they are at full penetration, so the dashed line represents the maximum you could install. So, for policymakers, you can say there's no problem promoting cavity wall insulation, or loft insulation now because we know it's going to be a part of 999 out of 1000 possible futures; whereas double-glazing, actually that's more a wait-and-see thing — you might just want to wait and see what happens (see the diagram on page 277).

So that's the type of stuff we do. This is a scenario-based one: so if we're looking at too many variables at once, then it doesn't really make sense to do the Monte Carlo thing. So, in that case, we just pick a few quite diverse but consistent narratives and then analyse those, and then say ok in a world where this and this and this is true, this would be the result.

The last question. In your field, for you, what would be the result, or the experiment, that has had the most significant impact, and why?

Well, as an academic, the paper I like the most — whether it's had the most significance or not — was a case-study within Newcastle. The co-author is at Newcastle University and he works very closely with the local authorities, and they were trying to understand their energy strategy through to

2050. So, we used this model and discussions to do that analysis; we incorporated all these sorts of uncertainty analyses with it, and the results were quite nice in terms of the elegant conclusions that you can present to a policymaker, because there were a few clear 'no regrets' policy decisions; now whether they've taken them and that's the impact it's had we can't really say, because we don't quite know what they're up to. But certainly, as the person writing the study, it was fairly clear that the result of it was a clear message, and that you were quite confident that it would work.

All of this stuff, as I say, they're normative models about the future, it's… Do you know the post-normal science literature? It's basically you're getting into this area where anyone can have a valid opinion about the future, about what the world's going to look like in 2050; so you kind of put your two cents' worth up there, but it would be foolish to say this is the way the world *is* going to be.

So, most of the research in your area is through case studies, or that's mostly the case?

Yes, it's pretty much all case studies. A lot of the motivation will come from the policy world — someone will want to know what does a future energy system look like in this specific location.

Chapter 15
A Conversation with Nilay Shah

Can you describe your field briefly?

My field is modelling of energy systems and process systems. So, process systems are for making products, whereas energy systems are for converting energy into different forms and eventually into a useful service.

And by-products are any type of products?

Primarily, we're working on things like polymers, health care products, some consumer goods, foods, things like that.

Present a paradigmatic example of a model in your field, describing it in terms that are accessible to non-experts.

Soya milk is quite a popular product nowadays, as a substitute for dairy milk. So, one of the questions is how do you reliably supply soya milk all year round when soya beans are harvested at different times in different parts of the world? How can you configure a supply chain that can, for example, reliably supply soya milk to Europe all year round, with good quality? Because if you just store the soya beans, then over time the quality of the soya milk that you'll produce from them won't be that good. So, can you set up a supply chain that looks at different parts of the world, for example, in the Northern and Southern hemispheres, where the harvest times are different, and coordinate that system so that you can always get good quality soya milk for people to drink. That's the kind of question we might look at — where are the best places to produce soya milk, and when would you have contracts with different suppliers to supply it so that you get a steady production for something which is quite a seasonal product in terms of when it's actually ready?

With the help of this example, could you explain why a model is needed? And could you describe what a model is? (In particular, what are the different stages or steps in the modelling process?)

A model is needed to explore the costs and benefits of alternative supply chains. For example, the costs of growing and processing soya might vary by region, as might the variability in the actual amounts produced and, of course, the harvest periods. Because the growing regions might be geographically distant from the demand regions, knowledge of transportation options and their cost becomes important. Eventually, you end up with too many trade-offs to try to find a good solution by hand and you need a model to explore these trade-offs and identify an economically efficient solution.

Is there a role for mathematics in this model, and if so, what is it?

Very much so. All of our models are ultimately mathematical models (i.e. a set of variables and equations). The way we build these models is essentially with some mathematical formalisms. So, we have our parameters which are the known data, like where can you grow soya beans and when in the year are they harvested? And on the other hand, how much demand is there, is it growing, is it shrinking? What are the logistics costs? And then our models have variables, for example, how much do you buy from different parts of the world and where do you ship it to? And then how much do you process every month, and how do you supply that? And then we have constraints, or equations, which say that you can't sell more than you produce, you can't produce more than the factories' capacity, you can't buy more than the farmer can produce, these kinds of constraints. Nearly always, our models are mathematical in nature.

Besides the mathematical formalism that is needed to write down the constraints or the equations, what else constitutes a model in your field?

Well for me, a model starts with about three different things: one is, what is it that you really want to know about the system? So, if it's the framing — what is your actual question? Because often people build models without really knowing what questions they want to answer.

So this is the part of, for instance, where you discuss with the producer, or the…

Let's say the stakeholders, yes. So a model might have a set of stakeholders — it might be a government that's interested in how to decarbonise its economy, or a company producing soya milk. So, you're trying to frame the questions properly before you start with any mathematics. There's always a temptation to dive into the mathematics too quickly. So, there's the conceptual framing of the model; then there's the actual conceptualisation of the model itself, and its elements… So, first it's just the questions, then you go into the model concepts — what are the elements of the model? what are the boundary conditions of the model, and what are the main parts of the real world that you are then going to idealise in some way in the model? Then there's, as you say, the mathematics, then the most important bit that again people almost forget about is the analysis. Because often you have to run it many times, do sensitivity analysis, do 'what if' analysis, see whether it's actually properly validated and performing as you'd expect it to perform. So, for me, there are these four steps: there's the framing, the conceptualisation, the actual mathematical modelling, and then the analysis. And often there's a bit of feedback from the analysis all the way to the framing, because often you realise that you've forgotten something.

We'll come back to the analysis later with questions regarding simulation. So, would you say the role of language, or qualitative analysis, is really important in modelling both at the beginning, a tool to frame the problem, and then at the end in order to…

In fact, in order to conceptualise the problem…in the framing you're just looking at the big question — what do we want this model to do? Then there's the model conceptualisation, which is what are the elements we want to include in the model? So, still here you're in a very semantic or qualitative place, and then it's only in the third stage that you move to a more mathematical description and in the analysis as well.

How important is the notation?

Probably not as important as framing the right question.

Dr. James Kierstead mentioned that as part of the modelling phase, one of the constituents is the data, the data is very important — such as the cost of fuel, or the cost of a gas boiler — I guess that's something also quite important in your field, which is quite connected?

Yes, I was thinking that in part of the mathematical description of the model will be all the data which surround the model. Some of these might be things that we know very well, some might be things that we roughly know, and some might be things that we just have to guess about in the future — like how much energy we're going to need, what sort of thing people might want to drink? But we might have some lessons from the past that can help us. So, yes, data is very important, and understanding how *well* you know different things is important.

Models are often said to represent a target system, for instance, in climate modelling your model represents the climate; is this the case in your models?

In some of our models, yes. For example, in urban energy systems, how a city works and how a city uses its energy, which is quite a complicated thing, but more simply, in how a building uses its energy, and how much heat you may need at different times of the year. It is trying to represent something physical, or something you think you can describe reasonably well. And again, for me, how accurate it is goes back to the problem of framing, because often in the framing of the questions you can actually decide how accurate your model needs to be to the system it's supposed to be representing.

What is the relation between the model and theory?

For me, theory has two roles to play. One is in helping you to conceptualise the model. Mere framing of the model cannot help us answer these questions — you can't go from that to getting to work on a computer. You've then got to break up that system that it's supposed to be describing into what are the elements; so some of the theory about how to model is to say what are the elements of the system, what are the interactions between them, how are you going to represent things like space, time, physical flows, like human behaviour, and so theory comes in strongly at that level of how you go from the frame to the conceptualisation. Then from the conceptualisation to the maths, there are a lot of mathematical theories, you know discrete time, continuous time, if it's a probabilistic system what kind of probabilistic framework will you use? So, there are a lot of theoretical frameworks you definitely have to follow; they're very system-dependent, I would say — so climate science has its own theories about convection and so on, these sorts of things they work on compared to what we work on.

So, if I may rephrase, you would say in the process of constructing your model you will have different inputs coming from different theories representing different aspects that have to be taken into account in your model?

Yes, exactly. That's where a good knowledge of physics or whatever is going to be important in our kinds of models, which are to some extent reflecting the physical systems.

What is the aim or use of the model in your field?

For us, it's nearly always to help somebody make a better decision about something. So, the one thing our kinds of models do, is decision support. We have done some work for companies on where to invest in assets; for other companies, how to manage risk. For example, in clinical trials, we've got lots of pharmaceutical companies trying to decide where they should start investing in production facilities if their products are successful, so we're managing risk. We are doing work with the Energy Technologies

Institute which would be about prioritising investments in R&D, and helping them make better decisions in energy technologies. So yes, a lot of it is coming down to how to make better decisions about investments or policy, that sort of thing.

So, does it mean that the outputs of your models are more qualitative, or directional, or quantitative?

No, they can be quite quantitative. So they may say, build a new factory on this city of this capacity. That's quite a quantitative answer. But in other cases, say, prioritise the development of this technology because it looks useful to the UK. So it's both. It can be very quantitative, yes.

Earlier, you mentioned computer simulation: what would be for you the relationship between computer simulations and the model?

For us, really, once you've built the model, that's a description of the system, then you use it in a simulation or optimisation framework, usually to run multiple case studies, to understand if I have these sets of assumptions or this set of inputs what are the model outputs and the model recommendations. So, you can explore what you think might be sensible results and answers for your original framing questions. So for us, simulation really means using the model, or running the model repeatedly with different inputs.

This is a way to test the robustness?

Yes, the robustness, to see if your hypotheses are accurate under different assumptions, or different data, like, let's say, future prices of fossil fuel which we don't know, so how do your models answer the change. That would be how we use simulations, if you want to use that word, it's repeatedly running the model with different assumptions.

Dr. James mentioned two different types of simulations: Monte Carlo-type simulations, which are a bit similar to what you have described, but also he mentioned scenario analysis — when you have less information. Is that something that you would also consider?

Yes. We would consider both of those as aiming at the same thing. They are just reflections of the information content of your system, so if you've got a lot of information about your system, but maybe still not perfect information, then you may well go for a Monte Carlo approach; but if you're trying to say something around geopolitics — like will we continue to get gas from Russia? — you might just look at that as a yes/no scenario, and look at that with sensitivity.

And in terms of the stage of the modelling, would you say that the simulation part would be the last one in the order, but that it can be back and forth…

Yes, because often you'll find at that point that you forgot something in the original framing of the problem. Nearly always. Because it's almost impossible to frame modelling problems correctly in the beginning. So, you'll nearly always go all the way back to the beginning, add some more questions…because as you start to get solutions, inevitably you'll discuss them with stakeholders and they'll say, 'oh yes now I remember it's worth also checking this question and that question.' So you nearly always go back and refine your model and you iterate quite often.

So, the stakeholder might be present at various stages — he's not simply waiting for the final output…

No. Often, the best thing with modelling is to build a rough and ready model, like a prototype model, and use it in engaging with the various stakeholders, and say, 'is this really what you're looking for, or is this behaving the way you want, is it giving the answers you want, how should we enhance it?' So, you should always build as simple a model as possible, especially early on, just to check that everything is on track.

What is a good model?

For me, a good model is one that is fit for the purpose it was framed for, and so it helps you answer the questions you want to answer, and it does so with reliability, and with some degree of validation. So, a good model isn't necessarily a hugely sophisticated model, it's a model that's fit for purpose, for that framing. That's why the framing is so important. So then it's got exactly the right kind of fidelity, and accuracy, and reliability for the framing question. It's fit for purpose for that set of questions that you're asking. So, if you're asking about future climate predictions, obviously that's going to be a very, very sophisticated model, but if you're asking some simple questions around approximately what kind of energy mix might we want to have in a city to meet some carbon targets, it may not have to be at the same level of sophistication.

Especially if the answer that will help the decision does not need to be very quantitative/precise.

Yes, exactly.

What has been the impact of the development of new technologies or tools? (e.g. in cosmology, telescopes, computer simulations for DNA, etc.)

The key developments have been in computer software for solving equations, optimisation models, and so on and in computer hardware. This allows us to develop more complex and sophisticated models covering more phenomena.

Moving a little away from the modelling stage to the notions of risk and uncertainty, how would you define uncertainty?

For us, uncertainty means that we don't know some of the things in our system, in our description, perfectly. So, they might be numbers that we know a range for; or it might be something about the behaviour of something in our system that we don't perfectly know, whether it's linear or nonlinear; or it might even be uncertainty in respect to the performance of some of the technologies that we might be trying to describe in our system, for example, because they aren't fully characterised yet or they don't exist in a material form. So, it manifests itself in different ways, but the way we often think about it in mathematics is, if you have a certain system or an element — normally our systems have elements in them — like James said, a boiler would be described by some parameters, like its capacity, its cost, its efficiency, and those are things that we can know quite well. But let's say we're thinking about how much cities in the future will have different technologies, like fuel cells, well we don't exactly know what a fuel cell will cost, what its capacity or efficiency will be, but we have some ranges. So, in a lot of our models, the way uncertainty eventually is described, is by comparing the certain case where we can describe something with some numbers precisely, to something where we're using some kind of probability distribution on the numbers — could be just a simple range, or it might be some kind of density function.

So, in this case you will test for robustness?

Yes.

And so uncertainty is more parameter uncertainty?

Yes. Well, that's not always the case. We often have models which have some structural uncertainty as well. A good example again is when we're modelling pharmaceutical systems, where we're looking at, we might have five or six products in clinical trials, and if the company is very lucky, all five might be successful, if the company is very unlucky, none will be successful, and on average it might be somewhere between, so it's a very structural uncertainty. Products will either be successful or not, there won't be anything in between. So, that's a much more structural uncertainty, which creates more of a scenario tree of possibilities.

So, it seems that the modelling process is more of an uncertainty taker — there is some uncertainty that will be in the modelling process and in order to handle it you might run different analyses or different scenarios?

You might run different analyses. Often, we will run two particular analyses: one is called uncertainty analysis, where we look at the structural uncertainties and parameter uncertainties, and we see how they propagate through the system. So, if we look at our outputs — the cost of energy, the cost of soya milk, let's say — we can look at the distribution of the output; and the second thing we can look at is a global sensitivity analysis, which still looks at the uncertainty of the output, but then it attributes that to some of the inputs or parameters, so 30% of the uncertainty in the output is due to the uncertainty in the future cost of natural gas, let's say, 20% is due to the uncertainty in the cost of boilers; so, you can actually attribute your variants backwards.

So, this will be in the case where the output, the help for the decision, is a quantitative variable and then you will have a distribution around this quantitative value, and the variance of the distribution, is this uncertainty?

Yes. And then you can have an analysis of the variance, basically to attribute to some inputs.

How would you define risk?

People often confuse risk and uncertainty. For us, uncertainty is just the characterisation of things we don't know perfectly and try to characterise as best as possible — the ranges, if they are continuous, or by yes/no probabilities, if they are discrete, like the success of clinical trials, but they are value neutral — they're just a distribution, they don't say whether the left is bad or the right is good or whatever. But when it comes to risk, we define risk as the probability, and sometimes also the magnitude, of an outcome that is unfavourable. So now, what we're doing is we're attributing some qualities to those distributions. We're now saying, this outcome distribution is telling us there's a 25% chance that you will lose money. So, going from uncertainty to risk is really about taking that distribution of outcomes and looking at how much of that is unfavourable. For us, risk is essentially about probabilities and magnitude of unfavourable outcomes.

So, let's say at the end of your modelling process you have, for a quantitative outcome, this distribution around the variance, and then you go meet with the stakeholder and you explain this, and then from this discussion you can quantify the risk?

Yes, because one of the things you can say is our model for this supply chain is saying that we should invest in a new soya milk factory in Brazil, say. So then you can say, 'if you do that, the expected net present value will be, say, $50M, however, the downside is that there's a 5% chance that the net present value will be 0 or negative, and maybe there's also a 5% chance that it might be more than $50M.' So, you're trying to present multiple outcomes, because no future is certain, so when we say we're trying to help make better decisions, what we're trying to do with risk is to quantify what is the probability of an unfavourable outcome. So, we try to couch that in language like 'your most likely outcome is a good net present value, but maybe there's a 5% chance you'll lose money,' and if they say that's unacceptable, we can go back to the model, this goes back to the framing, tightening up some of the risk factors, because you can embed risk factors, and bring them a more conservative solution. So, whether that's an energy solution, or the soya milk example, risk is a way of understanding these uncertainties — outcomes leading to unfavourable or undesirable results.

Now the last question. In your field, what is the experience or the result that has had the most significant impact on you? For instance, when you started you discovered this paper, or later on there was this publication that had really a big impact on you, in your field I mean.

I think in my field it's probably the fact that these IPCC reports, which are now having a lot of influence in terms of international climate change negotiations, are based on energy system models and integrated assessment models, which are similar to what we've been using, and some of the things that they're saying, which are that we can't meet climate change targets without some technologies that have a negative emissions factor, which we were also arguing a while ago, that's had a big impact. So, it's the fact that the international UN framework on climate change and the systems they use, which are for the fundamentally critical process of climate negotiations, were actually using quite similar models to what we use. That's had a big impact.

Chapter 16
A Conversation with Paul Embrechts

Can you describe your field briefly?

First and foremost, I am a mathematician with strong research and teaching interests in actuarial mathematics. At the Department of Mathematics of the ETH Zürich, I am responsible for the education of actuaries, i.e. insurance mathematicians. Based within a very strong research department, my teaching and research in this more applied area also mirrors that environment. Our position on the educational/research scale within the worldwide actuarial community is hence to be located at the more mathematical, quantitative end. So essentially, my current research field is Quantitative Risk Management (QRM) within insurance and finance, with a particular interest in the modelling of extreme events and in gaining a better understanding of concepts related to risk aggregation and risk diversification.

Present a paradigmatic example of a model in your field, describing it in terms that are accessible to non-experts.

The Gaussian Copula Model (GCM) entered Wall Street around 2000 as THE model for credit derivative markets everybody had been waiting for. Credit derivatives are financial products that can be used both as investment tools as well as hedges (safeguards) in financial markets centred on credit, i.e. bank loans. In particular, the GCM offered bankers a way to model interdependencies (also referred to as default correlation) within and between loan portfolios. By 2009, its original star-like (some even predicted Nobel level) status was downgraded to junk level when it became known as 'A recipe for disaster. The formula that killed Wall Street.' Its original aim was to come up with a pricing and hedging tool addressing the elusive notion of default correlation. The latter technical term refers to the problem corresponding to the crucial task of modelling and risk managing credit markets where several credit (i.e. loan) positions may turn sour at the same moment in time. Building on GCM, its subsequent relatively huge financial markets, with nominal volumes into trillions of dollars, ballooned. GCM had its origin in actuarial mathematics for the modelling of so-called "joint lifes," and indeed was used with success in survival analysis when addressing the joint survival times of patients in clinical trials or epidemiological studies like "the Broken Heart Syndrome," a temporary heart condition that is often brought on by stressful situations, such as the death of a loved one.

With the help of this example, could you explain why a model is needed? And could you describe what a model is? In doing so, please answer the following sub-questions:

What is a model in your field? What are the key steps in a modelling process?

In my area of research, models typically appear as output or consequences of mathematical theorems in answer to concrete questions from practice. For instance, copula models were introduced to answer questions related to the construction of multivariate models with given marginal distributions and with a specific dependence structure. The key steps consequently include: (1) listening to practitioners, trying to achieve a clear understanding of the practical question at hand, (2) the translation of discussions from step (1) into a clear scientific (in my case, most often mathematical) context, (3) model fine-tuning after initial feedback, (4) output production and verification on whether or not this output really answers the question posed; if yes, fine, if no, some further iterations restarting with (1) may have to take place.

What is the role of mathematics in modelling?

First of all, mathematics provides a language in which a given applied, quantitative question can be clearly understood and precisely formulated. A very important part of this (re)formulation is pointing out possible misunderstandings stemming from often vague original formulations. Then it provides conditions under which certain technical consequences (like the calculation of prices) can be derived. In the latter process, it also points at possible limitations of the mathematical tools provided. And though Goethe is quoted as having said: "Mathematicians are like Frenchmen: whatever you say, they translate into their own language and forthwith it is something entirely different," looking more closely at his writings on the topic reveals however how close he was to my interpretation above of the role of mathematics in modelling.

What constituents besides a mathematical formalism are part of a model?

A first, and in my mind crucial, constituent of the above process is a sufficient understanding on behalf of the mathematical modeller of the underlying field to which his/her techniques are to be applied. Further, any model needs to take into account to what extent it can be justified, or for that matter falsified, on the basis of (statistical) data. Increasingly, computer implementation has become a necessity, but at the same time it carries the risk that the end-user becomes more and more distant from the underlying mathematical modelling process. As a consequence, any truly important model should be based on a regular exchange or feedback from practical experience in the model's day-to-day use. One has to avoid as much as possible a black-box status of models crucial within the business environment. Finally, output from a model needs to be communicated to a broader, possibly none(or less)-technical audience. This puts clear standards of quality on model-output and model-documentation.

How important is notation in modelling?

Correct notation is absolutely critical, if only to safeguard clear understanding of the underlying practical issues.

What is the role of language in modelling? Are there qualitative aspects in modelling?

As already stressed above, the role of "language" is eminently important! All too often, a model is as good for business as its flexibility and ease of communication. This bears the risk that in this Darwinian contest, the easy-models become the surviving ones, as they are often perceived as "easy to communicate to management." And as a consequence, the more involved, possibly a better model, may fall out because communication of its structure was deemed "too difficult for management to grasp." One may replace in the statements above "management" with "end-user" depending on the application.

An excellent example from the realm of regulation for banking and insurance is the difficulty encountered, since the mid-1990s, in convincing industry and regulation to move from an if-measure, Value-at-Risk (VaR), to a much more informative what-if measure, Expected Shortfall (ES). The first risk measure, VaR, addresses the frequency with which future extreme events (losses) may occur, for instance, a 1-in-100-year or 1-in-10-day event; the second, ES, addresses the much more relevant severity question: what happens IF such a rare event takes place, how much does one stand to lose? The 2011 9.0 Tohoku earthquake off the coast of North-Eastern Japan was a 1-in-10,000-year event (a VaR-like frequency), much more important information at the time was (would have been) available in an ES-like severity estimate, "What is the expected tsunami height near the Sendai area coastal atomic reactors, GIVEN that such a (very) rare event takes (or would take) place?" And yet, for far too long VaR was defended (especially in the world of banking) as much easier to communicate. In my opinion, and also from practical experience, such statements yield a grossly unjustified perception of the intellect of "management" and "end-user"! Despite academic criticism on the use of VaR, going back to the mid-1990s (VaR was "born" around 1994), only now (2015) are financial regulation and industry warming up to the transition from VaR to ES. Nevertheless, in a recent study, the International Association of Insurance Supervisors (June 2015) released a document revealing that a majority of companies still favour VaR over ES on the basis of (1) easier to understand/communicate, (2) less difficult to calculate, and (3) the shareholder view: based on a limited liability structure of companies, once a company is insolvent (i.e. the rare VaR event happened), the extra information on "severity of insolvency" (encoded in ES) becomes less relevant to shareholders. Whereas worries (1) and (2) can easily be overcome (at least in time), worry (or better said, observation (3)) is more serious and very much hinges on a still prevalent mood on Wall Street of privatising gains versus socializing losses. I am not saying that a change from VaR to ES will come any way near to changing this attitude, but a more consistent reflection on "what if" rather than (just) "if," and this at all levels of a company, would be an important step in the right direction.

Models are often said to represent a target system (typically a selected part or aspect of the world). Does this characterisation describe what happens in your field? If so, could you say how a model represents its target? In other words, how do you understand the model–world interface?

As already alluded to above, the model-to-world interface has to be of feedback-type. The target system aspect is relevant. For instance, when an apple falls from a tree I am sitting under, I may be interested in the event that the apple falls on my head (for which Newtonian mechanics will do just fine). I may, however, be interested in the precise path that a given electron in the apple follows along its path downwards (in this case, quantum mechanics becomes relevant). The target has changed. Similarly, in the GCM example above, the target could be a very specific, small, isolated credit portfolio for which the GCM would work just fine. If, however, one wants to target credit products with nominal values in the trillions affecting broad financial markets, then the GCM is clearly a far too oversimplified model. So, a clear understanding of a model's target is of prime importance, as is the feedback along the evolutionary path of a model in its applications to the outside world. When do we leave the target space for which the model was originally designed? Are we sufficiently aware of and indeed do we understand the broader target regions? All too often, these important broader model issues are not, or at best insufficiently, addressed. One of the catch words along these lines is model robustness, a theme of increasing importance to regulators and industry alike.

What is the relation between a model and theory?

This question can be answered from numerous angles, my answer only gives one. I personally see, in the examples I have considered, a model as being embedded in a theory, the latter yielding a broad model environment in which models can be formulated and tested. For instance, in finance, the theory could be that of rational markets operating under some kind of Efficient Market Hypothesis, whereas a specific model would result in the Black–Scholes–Merton price for a European Call. Alternatively, one could be more interested in Behavioural Economic Theory with a model based on Prospect Theory. Hence, I see models as being grounded in a broader theory or theoretical framework. The usefulness of this "grounding" can be illustrated on the basis of the GCM: recall that at the height of the financial crisis, 2008, say, banks and some insurance companies suffered huge (multibillion) losses from their credit portfolios, like Credit Default Swaps and Collateralised Debt Obligations. By some, in a very naïve way, the GCM was blamed for the downfall: it did not work as expected, especially when markets were under stress. A Financial Times quote at the time stated, "Why did no one notice the formula's Achilles' heel?" The simple answer is: academics (mathematicians) who understood the broader theory within which the GCM was just one little model, understood perfectly why it would backfire in moments of stress. This message was published, communicated at conferences, voiced in discussions *well before* the crisis. No one in industry cared to listen: "As long as the music is playing, you've got to get up and dance" became the market participants' excuse later on. There does (and unfortunately most often) exist a fundamental dislocation in practice between a model being used and a basic understanding of the theory within which this model is scientifically embedded. Famous examples of such an embedding are, for instance, GPS tools within the theory of general relativity, or e-banking (in)security within cryptography and, hence, to a large degree, within the mathematical theory of prime numbers.

What is the aim/use of the model: e.g. learning/exploration, optimisation/exploitation?

In broad terms, models are there to help towards a better understanding of the (not just physical) world around us. In that sense, learning/

exploration/optimisation all play an important role along the path towards this ultimate goal. Models should never be advanced to the point where their formulation and analysis become THE ultimate stand-alone goals. One may perhaps do so momentarily, but ultimately we need to descend back to the level of practical understanding, i.e. to the level of 'exploitation' mentioned in this subsection's title.

In case you use computer simulations, what is the relationship between simulations and the model?

At the basic level, given a model, computer power allows us to obtain answers to questions by simulating numerous realisations of the same model and somehow counting the number of ways in which a certain event occurs. Like testing whether a coin is fair by tossing it over and over again and counting the frequency of heads, say. Physically proving that the coin is biased may be difficult, tossing it is not. Similarly, a model used in practice may be far too complicated to allow us to analytically calculate the occurrence probabilities of certain events, say, but once the model is written down and its parameters calibrated, one can simulate numerous realisations of the process, and it also becomes possible to gauge the consequences of (small) changes to the underlying parameters and/or assumptions. The latter is also referred to as stress testing and relates to the question of model robustness as mentioned above. It is no coincidence that this methodology is often referred to as a Monte Carlo simulation. Perhaps one of the key examples in everyday life is to be found in weather prediction which, perhaps contrary to common belief, has achieved significant quality progress over the recent decennia. All this due to an optimal symbiosis of model theory and model simulation. It is hard to think of any branch of applications where (model) simulation does not play a fundamental role. But note that, typically, a model is lurking in the background. At this point, I ought to drop the notion of Big Data, though a better name is Data Science. It should be clear that most models in use out there can be falsified given huge amounts of data, even the fair coin hypothesis. With a zillion observations, we start to discern even the smallest (presumably irrelevant) deviations from a hypothesised model. Modern data scientists even brag that models are things from the past, let the (big) data speak for itself. I personally am highly sceptical, just for the simple reason that 'big data' does not necessarily mean 'large information content'…the future will tell. I personally strongly believe that even in the Big Data circus, models have an important role to play. It is a bit like the clowns in a circus, they often bring out the truth, the real sentiments in a performance as an important yardstick or mirror of our own identities, our own strengths and weaknesses.

What has been the impact of the development of new technologies or tools in your field? (e.g. telescope in cosmology, etc.)

Clearly the advent of advanced computational tools combined with ever more powerful computers are no doubt key to my field.

What is a good model?

Here, I want to start with the often (mis)quoted statement of the statistician George E.P. Box: "All models are wrong, some are useful." The correct original statement from his 1976 publication in the *Journal of the American Statistical Association* (Vol. 71, No. 356, p. 792), addressing the issue of parsimony, reads as follows, "Since all models are wrong, the scientist cannot obtain a "correct" one by excessive elaboration. On the contrary, following William of Occam he should seek an economical description of natural phenomena. Just as the ability to devise simple but evocative models is the signature of the great scientist, so overelaboration and overparameterisation is often the mark of mediocrity." In the next paragraph, he continues with, "Since all models are wrong, the scientist must be alert to what is importantly wrong. It is inappropriate to be concerned about mice when there are tigers abroad." For those interested in the present volume, I strongly advise a careful (re)reading of Box's original paper. There is not much more that I can add beyond the fact that, mainly in the statistical, econometric, and insurance/finance literature, behind the following terminology, several tools and techniques are to be found: model selection, model validation, model adequacy, model uncertainty, model misspecification, model robustness… It would lead me too far to discuss (some of) these topics here more at length.

How would you define uncertainty? And how does the model help us understand uncertainty?

Benjamin Franklin is quoted as having said, "…in this world nothing can be said to be certain except death and taxes." For the moment disregarding both (death and taxes), everything in life is about uncertainty. This was scientifically, strongly stressed with the advent of Quantum Physics, to the extent that Albert Einstein exclaimed in disagreement, "God does not play dice with the universe!" At some point he added, "The Lord is subtle, but not malicious." This then provoked the following reply from Niels Bohr: "Einstein, stop telling God what to do." Whichever way one looks back at these early discussions between the scientific giants behind General Relativity Theory and Quantum Theory, modern science clearly puts randomness (hence, uncertainty) at the heart of it all. I personally do not want to enter into the philosophical discussions surrounding uncertainty (and risk, which we later will do), nor do I have the background to enter into a discussion on Quantum Theory. The interested reader may want to search for the plenary lecture I gave at the 30th International Congress of Actuaries in Washington D.C., April 2, 2014, with the title "Uncertainty." As stated in that lecture, for me uncertainty is about incomplete knowledge, an incompleteness I try to model to a high (though not complete) degree via a scientific (mathematical) edifice created for that purpose by A. N. Kolmogorov in 1933, when he wrote his path breaking *Grundbegriffe der Wahrscheinlichkeitsrechnung* (*Foundations of the Theory of Probability*). I realise that in doing so, I do dodge various important issues, which we later will come back to. I am convinced that by using Kolmogorov's triplet of "Sample Space," "Event Space," and "Probability Measure," we can do a lot in describing uncertainty in numerous applications. Within this, by now, classical theory of probability and statistics, one can (and has) achieve(d) considerable progress concerning the modelling of random phenomena. Let me just give a pedagogical example I often use in my introductory lectures on probability and statistics in order to discuss the notion of uncertainty and what to highlight when theory and models become important. Suppose you have a group of students divided up into two roughly equal subgroups. To students in Subgroup 1, you instruct to toss a coin 200 times at home and report the results (110010110001…) as these are produced. The second group goes home and writes down such numbers the way they think a fair coin produces such a sequence (111010010001…). Ask the students to write down the numbers, ordered as produced, on a card together with their names (but not their subgroup) on the back. I claim that by just looking at the card one can separate both groups almost perfectly (here the 200 plays a role). Indeed, one can prove (using the standard mathematical model for fair coin tossing) that Subgroup 1 (the coin tossing students) produce a longest sequence of 1s (heads, say), with very high probability, between 5 and 10. Subgroup 2 (the coin toss thinkers) will typically not come up with sequences of subsequent 1s (or 0s for that matter) of length more than 5. The precise mathematical model calculations are not so easy. One can, however, easily simulate this experiment and come to the same conclusion. I leave it to the reader to make the small step from this experiment to (high frequency) financial markets with stock prices moving up (1), down (−1), or no change (0). In a perfect random stock market, sequences of subsequent ups may turn out to be surprisingly long, just like for instance subsequent runs of black in roulette. On August 18, 1913, a staggering sequence of 26 successive blacks were observed in Monte Carlo. Model calculations can however explain that, in all the roulette games witnessed all over the world over a longer period of time, this is not such a rare event. A more interesting and telling (real) story is that of the statistician who beat online casino because the computer programme underlying the online roulette wheel simulations produced more switches from black to red and backwards than proper randomness allows for. After a short winning streak, our statistician was forbidden to play, and subsequently the computer programme was altered so as to generate spins of the wheel which resemble more closely truly random spins. This I find a compelling example of the power of modelling and I urge readers to consult on this and further examples from the realm of uncertainty in *Significance*, December 2013, 10(6), published jointly by the *American Statistical Association* and the *Royal Statistical Society*. D.J. Hand's *The Improbability Principle: Why Coincidences, Miracles and Rare Events Happen Every Day*, Scientific American, 2014, yields further food for thought. Finally, coming back to the stock market translation from long streaks of ups or downs for a coin or a roulette wheel to ups and downs on financial markets, entertaining texts on the topic carry titles like *A Random Walk Down Wall Street* (Malkiel) and *A Non-Random Walk Down Wall Street* (Lo and MacKinlay).

How would you define risk? And how does the model help us understand risk?

According to common usage, risk entails both uncertainty and exposure, i.e. possible consequences. In our textbook on *Quantitative Risk Management* (McNeil, Frey, Embrechts, Princeton University Press, Revised Edition, 2015), we give the following definition of risk: "Any event or action that may adversely affect an organisation's ability to achieve its objectives and execute its strategies." First of all, this rather restrictive definition only looks at the downside, and neglects the equally (if not more) important upside. Also, it implicitly links risk to measurable objectives and clearly formulated strategies. Reality allows for many shades of grey here: from purely analytical formulations to much more soft interpretations. And hence, also a multitude of names occur in the academic literature and relevant practice (especially within banking and insurance). For instance, in Asian countries the preferred word for risk is resilience with its more proactive interpretation. So, an important distinction is to be made between (1) aleatory risk, as the risk coming from random fluctuations, as in a coin-tossing-based game, and (2) epistemic risk, as the risk due to our incomplete understanding of a problem. For the first type of risk, probability and statistics offer excellent tools, for the second, the key advice is, "be aware of this not-full understanding and learn more." In my own experience, most real problems in practice are a combination of both types of risks, (1) and (2). A typical situation presents itself when one wants to price and hedge a complicated financial or insurance deal: besides the random fluctuations at the level of the underlying data used and model assumptions made, typically customer and market behaviour, the influence of changing accounting rules, the legal environment, and changes in the political landscape all play an important role. On the latter, politics, Donald Rumsfeld's statement, "There are known knowns. These are things we know that we know. There are known unknowns. That is to say, there are things that we know we don't know. But there are also unknown unknowns. There are things we don't know we don't know," at first may sound a bit hyperbolic, but at the time he made the statement it surely set risk and (un)certainty free for a broader public debate. On a more profound academic level, in his famous 1921 treatise on Risk and Uncertainty, Frank Knight equates risk to measurable uncertainty and reserves the term uncertainty for unmeasurable uncertainty. This distinction relates risk to objective probabilities, whereas uncertainty corresponds to subjective probabilities. Numerous discussions by many authors emerged, including John Maynard Keynes and Bruno de Finetti, the former a key figure in economics and finance, the latter taking centre stage in insurance. Leaving some of the more philosophical discussions aside, when we assume a middle-of-the-road definition of risk as any event where both uncertainty and exposure act together, then it becomes clear where modelling plays an important role. And this not only in attempts to quantify both ingredients, but also, and in my mind very importantly, in describing clear boundaries beyond which a given model output is not anymore justifiable. Here, we enter the increasingly important field of model uncertainty and robustness. For instance, if we want to calculate a risk measure of a given aggregate position of risk factors, but we have little information on the interdependence between the risk factors, it is better to give best–worst bounds on this risk measure and not try to force upon the end-user a single number. This kind of functional model uncertainty compounds the always existing parameter uncertainty, which is of a purely statistical nature. The story around the use of the Gaussian Copula Model mentioned at the beginning can perfectly be framed in this context: besides the parameter uncertainty in the underlying models for the default probabilities for the underlying credit-based positions, there was great functional (macroeconomic) uncertainty with respect to the interdependence between these credit positions, and this especially in moments of market stress. The latter occurred with catastrophic consequences for the world at large as soon as American house prices did something they never did before…they started to fall!

What is the role of stress testing and sensitivity analysis in the understanding of risk and/or uncertainty? (It might be helpful to clarify what stress testing and sensitivity analysis involve in your field).

A proper stress test and/or sensitivity analysis belongs to most applied problems I have worked on. This may be either disguised under the form of parameter and model uncertainty or explicitly demanded for, like for instance, in banking and insurance regulation. An example of the latter case consists, for instance, of the broadly publicised, and in some notable cases, not very successful stress tests by governments on the stability of their respective banking systems. A further example, this time from the insurance side, consists of the various stress tests demanded for by the regulators responsible for the (private) life insurance business. Such a test might include downturns in the stock market and even the calculation of the consequences on current balance sheets of past extreme events like the 1987 crash or the 2007–2009 financial crisis. Also relevant are the potential consequences of a pandemic.

What do you consider to be the work/result that has had the most significant impact on your field? And why?

I occasionally ask students or visitors passing by my office to single out from my rather extensive personal library on mathematics the couple of books that look most worn out. Almost invariably, the books chosen are William Feller's two volumes on *Probability Theory* and Walter Rudin's *Real and Complex Analysis*, as well as his *Functional Analysis*. The fact that these books go back to my days as a student of course helped in the wear and tear process, but even so. These two categories of mathematical texts exemplify my own attitude to the field of modelling. In Feller's books, one not only learns the more technical aspects of probabilistic thinking, but also the importance of intuition in the field of uncertainty. On the other hand, Rudin's texts are prime examples of the beauty and relevance of mathematical rigour. In my opinion, many (if not all) problems in practice need a combination of both skills: intuition and a basic understanding of the broader issues together with care for clear definitions and an in-depth understanding of the underlying conditions of the model within a more mathematical framework. The GCM once more is an example where this symbiosis of thoughts should have played more strongly. At least, we as mathematicians should foster both! It is impossible to single out a specific scientific contribution in my field of research with main impact on my work as such a choice so much depends on the type of (applied) problems I am currently working on. In my career, I have been uniquely blessed with working environments and intellectual impulses from numerous fields: starting with pure and applied mathematics as a student, early research in applied probability with actuarial applications, economics, finance, statistics, operations research, and industry experience from working together on concrete problems, to membership of boards of independent directors in the financial and insurance industry. It is this combination of experiences that has by far had the strongest impact on me as a(n academic) risk modeller. From these contacts and influences emerged the two (by now classical) books I co-authored with former students on *Modelling of Extremal Events for Insurance and Finance* (Springer, 1997) and *Quantitative Risk Management* (Princeton University Press, 2005, 2015). This educational path is difficult to copy; at all stages of life, uncertainty and risk, in the form of decisions and their consequences, have to be taken into account. I have been fortunate both at a personal as well as a professional level to have taken the right (or at least challenging, interesting) turns whenever the road ahead forked. By far the most important skills I learned are an openness of mind and an eagerness to learn both at a societal as well as academic level. And whenever I work on a truly applied problem, the words of Hamlet to Horatio are not far away: "There are more things in heaven and earth, Horatio, than are dreamt of in your philosophy."

Chapter 17
A Conversation with Ron Bates

Can you describe your field briefly?

My work involves the development and application of mathematical and statistical tools and methods to engineering problems for the robust design of industrial products and processes.

Present a paradigmatic example of a model in your field, describing it in terms that are accessible to non-experts.

A classic example of a structural mechanics problem is the beam problem. For a solid beam, restrained at one end, with a point load applied at the other end, what is the magnitude of deflection? This will depend on the shape of the beam, its material properties and the load applied. The beam itself could be modelled as a line (1D), a rectangle (2D), or a cuboid (3D). If the load is too great for the beam, it will break. Basic linear static physics models can predict the point of failure, but more sophisticated nonlinear physics models are needed to model what happens after that point.

A typical example of a model could be a Finite Element (FE) Analysis model. An FE model is used to give a numerical solution to a partial differential equation (PDE) problem that is specified over an arbitrary domain (usually defined by 2D or 3D geometry). A typical FE model is composed of 3D Computer-Aided Design (CAD) geometry, discretised into small elements by meshing, with appropriate physics (defined by the PDEs), material properties, boundary conditions, and forces applied to it. Taken as a whole, the model can be viewed as a system composed of a bounded domain (the geometry domain), a system of equations to be solved over that domain (physics domain), for example, the Heat Equation, Wave Equation, or in the case of Structural Mechanics, a plane strain formulation that describes how objects deform under loading, and a set of boundary conditions that link the geometry and physics (loads, restraints). The geometry domain can be 1D (no geometry), 2D (for example, a cross section), or 3D (a solid body). In general, there is no closed-form solution to such systems, so each case is solved numerically by the discretisation of the domain into Finite Elements and an iterative procedure to minimise an error function.

With the help of this example, could you explain why a model is needed? And could you describe what a model is?

These models are used to support and inform the design and development of new products and processes. These new products and processes can be thought of as systems that need to be designed and developed to satisfy a set of requirements. We develop models to test hypotheses, to evaluate physical effects over specific domains, to verify the operation of a system (by mimicry) and to validate that system against those requirements. A model in this context could be a way of evaluating the system to determine if it meets those requirements. For example, to design a beam that is below a certain target weight, that can support a prescribed load while not deflecting (bending) more than a certain amount, that costs below a certain amount to manufacture and that will last for a certain amount of time (fatigue strength). One or more models of the beam may be needed to test each of these criteria. An FE model of the beam could be used to estimate weight, strength, and deflection, a separate model may be needed to estimate manufacturing cost and fatigue strength.

What is the role of mathematics in modelling?

Mathematics is key: geometry for CAD, models of material properties, meshing (discretisation), representation of uncertainty, physics models (thermal, mechanical, aero), numerical solvers. Traditional CAD methods involve Boolean operations on primitive shapes, for example, subtracting a cylinder from a cube to create a plate with a hole in it. These solid objects then need to be discretised into small "finite elements" such as bricks, tetrahedra, or pyramids. In the limit, as these elements become smaller, they provide better approximations to the functions they represent. The challenge is to balance element count with numerical accuracy. Models of material properties are developed from testing specimens of materials and fitting models to the data.

What constituents besides a mathematical formalism are part of a model?

We could think of design knowledge, an appreciation of physical effects, test data, expert judgement. For example, the modeller needs to understand what the relevant or important physical phenomena are that need to be represented, and the mathematical model needs to adequately represent them. This depends on the context for the model and how it is to be used. It may require a model with several/many interacting sub-models, for example, a model of an engine compressor can be broken down into individual component models, each of which may itself have embedded models. These models need to be coupled and solved together. They will also need to be calibrated against test data to validate their use for a particular purpose or situation.

How important is notation in the modelling process?

Notation is important, especially when the model is composed of several sub-models that may be from different disciplines. Notation tends to be discipline-specific, which can lead to ambiguity. For example, i is generally used to denote $\sqrt{-1}$, but in electrical engineering, i is current, so j is used instead. Notation is also important for verification and validation of the model, to provide clarity to try and avoid errors, and to show that the model is the correct one for the task, and has been implemented correctly.

What is the role of language in modelling? Are there qualitative aspects in modelling?

In expressions of uncertainty about the model, in descriptions of the model itself, how it will be solved, the context in which it is valid, the decisions to be made as a result of evaluating the model. Language describes the interface between the model and the physical world. For example, in creating 3D geometry models, one needs to describe primitive shapes and their intersections (Boolean operations) and extrusions. In the context of the use of that shape, these become blends, fillets, flanges, etc.

Language is also important for data reduction, and for the embodiment of the mathematical model in computer code (itself a language). Particularly in geometry, something like Generative Modelling Language can be used to compactly represent very complex shapes with generating functions, rather than an explicit description.

Models are often said to represent a target system (typically a selected part or aspect of the world). Does this characterisation describe what happens in your field? If so, could you say how a model represents its target? In other words, how do you understand the model–world interface?

Yes. Models are used to mimic specific behaviour of the target system. Systems are composed of subsystems which themselves are bounded systems. Thus, system behaviour may itself be decomposed. There is therefore a hierarchy of models (Consider building a model of a piano. One may start with individual keys that produce notes. Each model of a key could be composed of sub-models for the mechanical operation (feel) and vibration models (richness of sound) and so on) which can all be related to a corresponding hierarchy of test data, up to field performance data.

What is the relation between a model and theory?

The model is an expression of the theory. Models are used to represent systems at a theoretical level and to understand system behaviour, to give us confidence in the theory. In the beam example, there may be a new theory that a hollow beam, or an I-section beam, could be used for a particular application. In this case, the Finite Element Analysis model would be used to evaluate the new design for weight, strength, and deflection for a given loading condition (boundary condition). For a more complex example, consider the need to evaluate a new design (theory) for the piano. Historically, one would build a physical prototype to evaluate the design, but one could also build a mathematical model. The piano keys are geometric/mechanical, a model for vibrating strings needs to be developed, and so on. As the problem is broken down, each subsystem (or sub-subsystem), presents its own modelling challenges and a balance needs to be struck between the cost of physical prototyping and that of mathematical modelling and the risk that whichever route is chosen, the model is a true enough representation.

What is the aim/use of the model: e.g. learning/exploration, optimisation/exploitation?

Ultimately, it is part of the product design process and is used to make decisions on the design, manufacture, and use of the final product. Models are used for calibration (against physical test), as part of the product certification process: for Sensitivity studies, for Design Trade studies, for Robust Design, for Robust Optimisation. Models are used at all stages of the Product Development Process (Preliminary/concept, detailed, calibration).

In case you use computer simulations, what is the relationship between simulations and the model?

Simulations are evaluations of the model, both static (equilibrium) and dynamic (evolution of system response over time, with time-varying model components). The model is usually an FE model comprising partial differential equations defined over a domain (usually 2D or 3D geometry). Simulation is then the act of evaluating the model over the domain. The simulation solves the model numerically to determine the responses of interest. Where the domain represents the solid model, these responses are typically thermomechanical in nature (e.g. determining stress or temperature over the whole solid domain, or displacement of the solid). Alternatively, one can simulate fluid flow in the space around the solid (Computational Fluid Dynamics, CFD).

What is the role of stress testing and/or sensitivity analysis in modelling? (It might be useful to be precise about what sensitivity analysis and stress testing mean in your field.)

One needs to be careful here to distinguish between stress testing a model (testing a model to beyond its limits) and stress analysis (determining the stress/strain in a solid object subject to loads and restraints (boundary conditions)).

By performing multiple computer simulations, for example, as part of a structured experiment, one can build relationships between model inputs and model outputs. Sensitivity analysis seeks to understand and rank the importance of the model inputs to each of the model outputs. For example, where would be the best place to add (remove) material from the beam to increase stiffness and reduce weight? Stress testing a simulation model could involve determining the limits to which model inputs can be varied before the model fails to solve (converge), or fails to give valid results. In the latter case, the model could be tested beyond the validity of the underlying mathematical model if the yield stress of the material is exceeded. If that were to happen in real life, then the beam would buckle or crack, but if those physical mechanisms are not represented in the model, it will not behave realistically. A simpler example would be physical contact. If the beam is bent so far that it would hit a nearby object (wall, floor, etc.), the simulation model would need to know the rules for contact, otherwise the beam might pass through the wall rather than being stopped by a collision.

What is a good model?

In this context, good means "fit for purpose." A good model is one that models the <u>relevant</u> features of a system and accurately mimics the behaviour of those system features that are under investigation (i.e. the features that have been shown to be critical). So it all depends on the purpose of the model. If the beam model needs to represent certain failure modes such as buckling or cracking, or needs to model contact with another solid object, but does not, then it is not a good model.

How would you define uncertainty? And how does the model help us understand uncertainty?

To bring further insight, beyond the obvious, uncertainty can be classified into uncertainty due to randomness (aleatory), due to lack of knowledge (epistemic), and due to not following best practice (incompetence/ineptitude). In measurement systems, uncertainty can be defined as a lack of precision. Going back to the beam example, there may be uncertainty in the exact dimensions of the beam geometry (epistemic — better measurement systems could help resolve this), there may be variation in temperature that affects material properties (aleatory — outside the designer's control), the modeller may be using an inappropriate model, for example, a Finite Element mesh of the geometry that is too coarse to accurately model the beam behaviour (incompetence).

The model helps us put uncertainty into context (uncertainty in model inputs, model discrepancy (with the real "physical" system), model use (context)), but in order to use this model, we need to define uncertainty in a stochastic sense as variation in model inputs, model parameters, and model use.

Considering uncertainty as a lack of knowledge, a division can be made between controllable (where risk-based models or probabilistic models can help) and uncontrollable (for example, change of policy or divided opinion among experts) uncertainty. For these uncontrollable "deep uncertainties" that cannot be defined in a stochastic sense, we need to develop representative model scenarios and evaluate each one separately. For example, the basic beam model may assume that the material properties are homogeneous. We know in practice this is not true, but it is not possible to know the true material properties at every location in the solid model, or how they might vary. In this case, one could develop sets of plausible material property distributions that are equally likely and solve the model for each one. The point is that we may not be able to build representative models of variation in the model parameters (the inputs to the model, the parameters of the model, the boundary conditions, the material properties, etc.), which leads to a more subjective/qualitative assessment of the likely or reasonable behaviour of these quantities.

How would you define risk? And how does the model help us understand risk?

Risk is the probability of a hazard occurring combined with its potential impact and how controllable it is. Models can help us understand (bound) the magnitude of each element of risk (probability, loss, controllability).

What has been the impact of the development of new technologies and tools in modelling in your field? (e.g. telescope in cosmology…)

The engineering analysis field is dominated by FE analysis, in both the structural (FEA) and aerodynamic (CFD) fields. Computational Fluid Dynamics models are more complex, due to the difficulties in modelling turbulence. However, both computational modelling tools provide massive capability to design complex engineering products. The ability to accurately measure manufactured geometry (laser scanning) provides the opportunity to replace "ideal" design geometry with "as manufactured" geometry to build more accurate models and gain further insight into the effects of manufacturing variation and wear on performance.

For you, what is the experience/result in your field that has had the most significant impact? And why?

This will have to be large-scale engineering failures (failures of large engineering projects, products, or systems that have had a large impact) due to "emergent behaviour" of complex systems (e.g. the Tacoma Narrows Bridge, which collapsed in 1940; the Millennium Bridge, nicknamed the "Wobbly Bridge," which had to be closed for nearly two years to eliminate the swaying motion). Failures such as these have huge publicity, and often lead to better understanding of new modes (ways) of failure, leading to better models, because the inadequacy of current models is exposed.

Chapter 18
A Conversation with Simon Dietz

Can you describe your field briefly?

I'm an economist, and most of my research is on environmental issues, in particular, climate change.

Present a paradigmatic example of a model in your field, describing it in terms that are accessible to non-experts.

The models in my field are often called 'Integrated assessment models'; that's because they integrate the insights of physical and/or natural sciences on the one hand, and social sciences on the other, usually economics. So the easiest example is an economic model of climate change: you've got a model of the economy, of varying degrees of complexity, and that is linked, through emissions of greenhouse gases, to a model of the climate system, and there may be another link between the climate system and the economy again, to close the loop. These links can be regarded in themselves as models, for example, to link the climate system with the economy, one essentially builds a model of environmental, social, and economic damage from changing physical conditions. A wide range of economic models can be coupled with a wide range of climate models, with computational constraints often being the limiting factor. The economic models can range in complexity from just a few equations describing economic growth and greenhouse gas emissions globally on aggregate, to hundreds if not thousands of equations doing this for different sectors of the economy and different parts of the world. Similarly, climate models range in complexity from just a few equations describing how the global mean temperature responds to the atmospheric concentration of greenhouse gases, to models with huge spatial and temporal detail.

So if I understand correctly, in your field a model would have a mathematical structure, your input would be different variables or parameters characterising some features...

Yes, either initial conditions, or assumed relationships.

And then you put these things into the quantitative framework and get some quantitative outputs, representing different features of the system that you wanted to model?

Yes, that's right.

So, with respect to this example, can you explain why a model is needed? Or the potential use of the model? Or potential outputs?

The following observations may not be separate, but two things come to mind. One is that we build models because we want to understand reality — so we create formalised abstractions of reality just to try to understand how the system works; and secondly, because we want to undertake experiments with these models, which is not possible with Nature. So, in the example of climate change, as I'm sure you're aware, we run these models to see what climate change is going to perhaps do to the global economy, regional economies, in the future — we don't observe that now, so we have to model it in order to try to understand it. It's been termed 'numerical control,' as distinct from 'experimental control,' the point being that we only have one climate system and we are not in a position to do scientific experiments with it to see how it responds to greenhouse gas emissions, so we build numerical models — integrated assessment models — in order to construct simulations of experiments.

What is the role of quantitative modelling, of mathematics, in the model?

It varies. The first thing to say is that within my field there are analytical models (yielding algebraic solutions, without depending on attaching quantities to the parameters and variables of the model — that give you qualitative insights into how the system works), and there are quantitative models. In other words, an analytical model will describe relationships between elements of the climate–economic system qualitatively, and the hope is you can say something about how, for example, the process of economic growth drives the global temperature, without having to specifically parameterise it and therefore open up to uncertainties. But the trouble is, there are limits to what you can say qualitatively; very often, the relationships you get are unclear and depend on the numbers, so that's where quantitative modelling becomes valuable.

This question of the role of mathematics is derivative of what the models are used for. If you're using the model to understand the system, then mathematics helps formalise and make precise a researcher's understanding of the system, so that other people can look at it and say that 'I understand it differently, and this is how I'm going to do it.' In the non-quantitative and non-mathematical social sciences, this is something that, or so I believe, they grapple with all the time, and they use jargon and language to try to make precise what they're saying, and differentiate themselves from other researchers. But we use mathematics.

On the other hand, when the models are being used to make forecasts — they are actually making some kind of claim to predict what the system will be like at some point in the future. Then I suppose the mathematics, as well as making things more precise, is intended to relate the model to real outcomes, typically numerical quantities like temperature, or economic outputs. It seems to me there's a subtle difference between those uses of math, but maybe not.

So, would you say mathematics is a common language?

Yes, I think it does enable us to precisely understand what our peers are doing, at least *within* disciplines or fields. I've found, for instance, that economists and physicists can engage in clear dialogue on their models, because they share many mathematical similarities (modern economics borrowing a lot from [old] physics). Conversely, I think this is why textual jargon is so prevalent in some non-mathematical social sciences, like perhaps, sociology. But it's important to realise that mathematics is also a barrier to entry for new-comers. So it's not an unalloyed good.

And as well as mathematical tools or equations, quantitative…what else do you think plays a huge role in modelling in your field? For instance, other forms of formalisms, or language?

I'm not sure I understand this question completely. Can you give some examples of what you're thinking about?

For instance, in certain fields or in economics, you have a huge difference between what people mean by risk and what people mean by uncertainty. And of course, this will have implications mathematically-speaking, or quantitatively-speaking, on how it is formalised, but the vocabulary that is used, the language, introducing new words or terminology, has an impact on the modelling, if that makes sense?

Yes, in the sense that these words are manifestations of new ideas, I suppose. So, someone who just talks about risk in a general sense is thinking about it in a different way to someone who distinguishes between risk and uncertainty, and that second person has a more nuanced understanding of risk than the first person. On that basis, for sure language is important insofar as it represents different ideas that people bring to their modelling. There are different schools of thought, in the area of integrated assessment, which is a somewhat interdisciplinary area, there are bigger battles, competing over how to understand these systems.

How important is notation?

In my experience, clear notation is invaluable and it is possible to seriously misunderstand a mathematical argument just by dint of the use of poor notation. On the other hand, occasionally notation goes into needless detail. I suppose it's a bit of an art rather than a science.

So, with this integrated approach involving different fields, are these big battles over assumptions, inputs to your model, the modelling itself?

From a modelling point of view, they are usually about structures, as opposed to parameters. There are, of course, vehement disagreements in say economics about some parameters such as the discount rate famously; but these are narrower arguments than the arguments between completely different theories of how the economy works — like the Keynesians versus the neo-classical economists, and they have different model structures that they consequently bring to bear.

So, if I use an analogy that is completely different, if you're baking a cake, it's about whether you use brown sugar or white sugar, rather than the quantity?

That's a reasonable analogy, yes.

A large part of models are said to represent the target system, to represent an aspect of the world. Does this apply to your field — I think you answered this already.

Yes.

How does the model represent the target in this case? Do you have special features you like to have in your model to represent the system — you simplify the system and identify certain features that you think are the key features that your model should represent?

Yes. The first thing to say is that the world of integrated assessment modelling is incremental to the much larger bodies of work in all whole disciplines that it stands on. As far as the economics parts of these models are concerned, they are not inventing a model, they are taking a model which has a much longer heritage in economics that has been used for other purposes and they are applying it to the climate problem. Going back, then, further into the heritage of these models, of course, how well they represent the target tended to be judged on how well they were able to represent some key stylised facts about the economy.

So that's how you choose which model to use?

Yes, exactly; and how the evolution of these models was dictated by, for example, saying that this particular model doesn't reproduce some key stylised fact and therefore we need to improve it, add more structure, change the structure so that it does. And similarly, the physics parts of the climate–economy models have a heritage in work in physics, which I don't understand, it's not my area.

What has been the impact of the development of new tools and new technologies in the modelling in your field? (e.g. as telescope in cosmology, etc…)

To some degree, advances in computer power and in things like optimisation routines have improved work in my field, but actually, in the economics of climate change, I don't think computer power is the limiting factor — there is a basic lack of empirical evidence, from the 'field' broadly speaking, of the impacts of climate change and of the consequences of policy changes to reduce greenhouse gas emissions, and this means the enormous spatial and temporal details of the sort you would see in climate physics are not really justified and simpler models are better, because they cannot be said to be less accurate, and they are usually easier to understand. Where new technologies are enabling big advances is around the fringes of what I do, personally. For example, remote sensing has greatly improved our understanding of the prevalence of conventional air pollution (i.e. soot, sulphur, nitrogen, fine particles). This relates to climate change because burning fossil fuels, particularly coal, produces both CO_2 and these conventional air pollutants. Another example is the use of Geographical Information Systems software and the Internet to improve economists' ability to understand people's valuations of the natural environment. To give a specific example, we used to just ask people what they would be willing to pay to protect, say, a wetland area, and maybe we could show them a photograph or two of that wetland and how it would be degraded without them paying for protection, but now we can use GIS to simulate much more accurately what we have and what would change.

How do you communicate with the climate scientists?

It has happened in various ways in this field. Some of the models have been built by individuals or small groups of people whose expertise does not span the range of skills required, and they've obviously gone and talked to and read work in the fields they don't have mastery of in order to build their models. In other cases, you have teams of researchers who cover everything, basically, and who work together on the different components.

And do you notice any obvious difficulties in collaborating with people so far away in terms of expertise?

Well, of course sometimes these groups of people talk past each other and it can be difficult to get everyone to fix on talking about the same thing. That's one difficulty. There's obviously a certain strangeness of concepts which confronts physicists trying to understand economics and vice versa. So these would be natural difficulties.

Does mathematical formulation help when doing interdisciplinary research?

Yes and no. I would say it depends hugely on what disciplines are being connected. As I said, I think it helps economists work with physicists because there is a shared language. But it is going to make it hard for economists to work with qualitative social scientists.

If I may go off on a slight tangent: many of the models in my field are not computationally demanding models, which means that it's quite easy to have a discussion about what's in these models and easy for people to pick them up and use them, and for the models consequently to diffuse around the research community. But at the computationally intensive end of our field, you do have this problem that you have this concentration of a small number of models in particular places, maintained perhaps on large computers, and then you get problems of understanding what's in these models. I'm probably going forward... Just thinking about communication.

It's interesting to see what could be the role of the computer...instead of being something that helps in solving, it could sometimes bring additional difficulty in communication.

Yes, all else being equal, I think they do make communication more difficult — insofar as more complex models are more difficult to communicate about, clearly.

In your field, what would you say is the relationship, or the difference, or the link between model and theory?

I think there's a one-to-one correspondence in cases where one particular theory purports to describe a whole system. But the difficulty in my area is that because you're trying to integrate insights between the sciences and social sciences, there really isn't a theory that's going to cover the whole thing. And that means that while you can build a model which is more or less in one-to-one correspondence with the theory for part of the system, that theory is going to have nothing to say about other parts of the system. And once you're in that world, where you've got different theories informing different components of your model, it's no longer clear to me whether there are interactions, in the sense that your model no longer corresponds to your theory for the bit that you think it does, because of the way it interacts with the other parts.

So, if I try to rephrase it, you're saying that in your field, a model will encompass different theories, and a model if you restrict it to a particular aspect does not necessarily correspond to the theory in this particular aspect because of the interactions.

Yes, I think so. The models transcend theories, because well-formed, well-rounded theories tend to concern themselves with a particular discipline, I don't really know of a theory of everything which is capable of dealing with the natural world and the social world.

We touched on this earlier, the aim or use of the model, when you talked of forecasting: in your field, is it more about learning, understanding the system, optimising, exploiting something?

So, Keith Beven draws a distinction — at least, I've seen it in his book (*Environmental Modelling: An Uncertain Future*, 2008, Routledge, London)... I'm sure others have drawn a similar distinction — between diagnostic models and prognostic models. Models used for understanding, just built to understand a system, and models used to forecast a system. And it's always struck me in my field that the boundaries between these things are blurred, because it's rare that the models I'm talking about have been built purely to understand; almost all the work that is done with them has some explicit or implicit claim to forecasting a future quantity. On the other hand, I don't think any of the researchers concerned would be comfortable with saying that they're just doing forecasting. So, it's a curious mixture, too. That may cover part of your question at least.

And what would you qualify as a good model in your field? Or is there any good model?

Yes, sure there are. There's obviously this well-known issue of model verification that concerns any kind of model of climate change — whether it be a physical model or a physical and economic model. We're making predictions in the future of a system which is going to be subject to a set of boundary conditions which have not been experienced before, so we can't exactly say whether its predictions are right until we actually arrive there. We can, nonetheless, attempt to verify components of the system, and so a good model would be one for which parts of it can be verified, certain relationships that it claims; it doesn't answer the question as to whether the sum of the parts makes sense, but a model for which individual parts don't make sense can only, at best, be right for the wrong reasons...so that doesn't seem like a good model.

The other way in which models in my area can be deemed to be useful is whether they have yielded valuable qualitative insights, whether they've helped us understand the nature of the problem better. So, take for example models that economists have built of climate change, which have contributed hugely to our understanding of the intergenerational trade-off, the sensitivity or the value of cutting greenhouse gas emissions to how much we care about future generations.

For this particular aspect, is it because of the modelling process — while people were modelling they realised this was important? Or was it because of the output of the model?

I think the original builders of these models knew that this trade-off existed, in broad terms, so it wasn't something that was just discovered as a result of building a model. However, they helped to make it precise and stark for people to see. So they made a valuable contribution in that regard. And there are many examples, that's just one.

Coming back to the computer aspect, if in your field people use computer simulations, or use computers to create their models, what is the relationship between these simulations and the model? For instance, you have your model, which is abstract, mathematical, and then you test your model using simulations? Or you don't even have the abstract thing, and then simulations are the model?

No, it's the former. If you're trying to build a model to understand the system, then you simulate when you can't solve the model analytically. And obviously if you're trying to forecast from a model, then you simulate.

So, it's a tool to use the model, or test the model, or to get outputs from the model?

Yes.

But it's something that comes, somehow, after the modelling phase? It's not independent, because it's part of the verification of the model...

I think it does. I think there's a sense in which you sit down with your pen or your paper or your computer screen and you write down the model. But at some point you press 'go,' and you just have a look and see what happens. So that suggests that...

You go back and look at the model...

Yes, of course, and you might think 'well that looks strange or wrong' or...

Moving away from models now: the next two questions are about uncertainty and risk. First, uncertainty: how would you define uncertainty?

I would think of the future as comprising at least two states; I would then... Well, a colloquial definition of uncertainty would be just that. A more precise definition of uncertainty in contrast to risk would be one in which I could not assign unique probabilities to the states of the future that exist.

So, you managed to define both risk and uncertainty at the same time!

I did, yes.

How does the model in your field help to understand uncertainty and/or risk?

They can be used for both purposes. These models can be... You can put probability distributions over parameters of these models, and then run these models, and therefore, the model would help us understand risk in that regard. You can do more simple things with these models, which are perhaps better described as using the model to understand uncertainty such as simple sensitivity analysis scenario-based modelling; or even more sophisticated understandings of uncertainty, where you look at different probability distributions.

So, for climate change, for instance, you could look at what will happen if the temperature was going like this, or...

Yes, exactly. So you could generate a probability distribution of outcomes and that would be some kind of conception of climate risk; or you could have different scenarios of what happens but you don't give them probabilities and that might be a simple way to characterise uncertainty.

And would you say, in your field, you would look more at things with a risk angle or with an uncertainty angle?

Well, there's a tendency in economics to think about things in terms of risk, just because the economics of risk is very well developed using the power of expectational operators. But there is a growing awareness of uncertainty contrasted with risk, and an increasing application of theories of decision-making under uncertainty to environmental problems, in general, and climate, in particular.

And these notions of risk and uncertainty, would you say they are more related to the risk and uncertainty of the inputs of a given model, or even to the model itself?

A typical approach is to take the structure of the models as being given and then to look at input uncertainty and how it propagates to output uncertainty. But I suppose, of course, there is also an uncertainty about the structure of the model, which is more usually observed in the differences between different people's work, and it may less frequently be thought of as uncertainty in the statistical sense, more like a battle of ideas, where somebody is eventually going to be proved right.

Can I just go back... I suppose the one thing is that when people do model inter-comparisons, that's a way to try to deal with uncertainty about structures of models. So you take a group of models which have different structures, and you try as far as possible to impose on them consistent sets of assumptions about inputs, and then see how they differ, then you're obviously trying to measure the structural uncertainty, let's call it, as well as input uncertainty — because there's no way you can impose uniformity on the inputs I think.

And now a [slightly] more personal question: in your life as a researcher in your field, what is the result, or experience, that has had the most significant impact, and why?

Well, as far as impact on me is concerned, that's actually quite simple to answer: when, by a mixture of design and chance, I finished my doctorate and went to the Treasury to work for Nick Stern on his review of climate change, that had a transformative effect on my career, which is still felt today. If I hadn't done that, it would have taken a very different route, both in terms of topics and success I think. And that's not particularly by design, so that's maybe one aspect of your question.

What I had more in mind is, if you had to pick one piece of research — whether very old or very recent, it can be yours or it can be somebody else's — which you think is a major breakthrough in the field, which one would it be, and why? Let's say you have to go to a desert island, and you need to take one piece of research…

That's very interesting. Can I have a few? Two or three? Well, the original work in the field that I've been talking about which was largely done by Bill Nordhaus (especially his book with J. Boyer, *Warming the World*, MIT Press, 2000) at Yale has obviously been very important, it didn't exist, this field, essentially Nordhaus…

Why, because he was the first one to do the seminal work in the field?

For a few reasons I suppose. He was one of very few people working on climate change in the late 1980s and early 1990s in economics. So, part of it was the sheer originality of it, and the second thing is that he's very good, and so he thought about it in a very elegant and useful way, which has endured while other people's research has been largely forgotten, which was quite similar in some ways. So, he would have to take a measure of credit.

Then, actually, after that, … I've been influenced a lot by people who worry about uncertainty in climate forecasting. I have indirectly been influenced a lot by Lenny Smith (especially his 2007 book, *Chaos: A Very Short Introduction*, Oxford Press, or his 2007 paper with D. Stainforth, M.R. Allen, and E. Tredger, "Confidence, Uncertainty and Decision-Support Relevance in Climate Predictions," *Philosophical Transactions of the Royal Society A: Mathematical, Physical and Engineering Sciences*, 365 (1857) 2145–2161), it's just that it took me years to figure out how what he was talking about might be sensibly structured in an economic model. I originally assumed that I might work quite closely with him, but it turned out there's such a huge difference between the work that he does on the predictive skill of models in the statistical frame and the way that economists think about decision-making under uncertainty. But I've got to give him credit for making me think deeply about issues with uncertainty.

And then, I suppose Martin Weizman's work on fat tails and structural uncertainty (for instance, his recent book with Gernot Wagner, *Climate Shock: The Economics Consequences of a Hotter Planet*, 2015, Princeton University Press), that's also work that I cite a lot and has been influential. So, I'd probably have to take those three.

Chapter 19

A Conversation with Stephan Hartmann

Can you describe your field briefly?

I am a mathematical philosopher, which means that I use mathematical and computational methods to solve philosophical problems. Having a background in physics, I am especially interested in applying modelling and simulation methods, which happen to be enormously powerful tools every philosopher should know about.

Present a paradigmatic example of a model in your field, describing it in terms that are accessible to non-experts.

One of my current research interests concerns reasoning and argumentation in science. I want to understand how scientists reason and argue, which reasoning and argumentation schemes (or types) they are using, and how good and convincing these reasoning and argumentation schemes are. It turns out that modelling methods are of much help here. Consider the following example. Physicists want to convince us of their favourite theories. To do so, it is natural to point to the successful empirical predictions the theory makes. But sometimes it is impossible to make empirical predictions, and sometimes there are no empirical data at all. So, what can we say about such theories? Are they scientific theories at all?

String theory is a case in point. The theory is highly ambitious. It pretends to account for all "fundamental" forces and it has several other desirable features. But it cannot be tested empirically. Not now, and perhaps never. So, how can one argue that string theory is a scientific theory, that it tells us something about our world, and that it is not just a nice mathematical formalism? Here, the 'no alternatives argument' (NAA) comes in. Scientists argue as follows: 'Look, string theory satisfies a number of desirable conditions: it unifies all "fundamental" forces, it connects nicely with established quantum field theories and even with general relativity, etc. And (now comes the important second premise of the argument), despite a lot of effort, and although we scientists are so smart, we have not yet found an alternative to string theory that also has these nice features. Therefore, we have one reason for the truth of string theory.' This is the NAA. The question for the philosopher of science then is whether this is a good argument. More specifically, we have to explore under which conditions the NAA is a good argument. To do so, we have to construct a model of the NAA.

Would you say that, in your field, it is about convincing? You gave an example in physics, but you might apply a similar modelling for other things, so for example, a politician trying to argue in favour of something? It could be the same pattern, right? The fact that it is applied to modelling in physics is just a particular coincidence?

Yes. The goal of an argument is to convince someone else (or perhaps also oneself), and different argument types are used in different contexts. The NAA, for example, is also very popular in political contexts. The older ones among us may remember Margaret Thatcher who famously argued that there is no alternative to economic liberalism, and Angela Merkel argued more recently that there is no alternative to certain fiscal policies for Europe. Therefore, what we have must be good, right? These are NAAs, and it is not at all clear whether they are convincing. Can we really be sure that there is no alternative, just because we haven't found one yet? This is where the philosopher of reason and argumentation comes in. We have to identify the relevant argument type and study if and when it works.

So, that's perfectly clear. With the help of this particular example, can you explain why a model is needed? And, in particular, it would be good to have a description of what the model is, and the various stages in the modelling process.

In order to address the question of if and when NAAs are good arguments, one needs a normative framework. The situation is actually quite similar to the situation in science. If we want to model systems like the planetary system or the hydrogen atom, we often choose a modelling framework — e.g. Newtonian mechanics or quantum mechanics — first, and then construct a model within this framework. Here, the framework constrains the modelling assumptions; it is something like the *stage* on which the action takes place. In our case, the framework has to be normative, i.e. it has to contain standards for the acceptability of an argument. In our analysis of the NAA, we choose a probabilistic framework known as *Bayesianism*. The idea is simple. We assume that scientists attach a probability value to a certain theory or hypothesis. This probability reflects how strongly a scientist (or a group of scientists) believes that the theory in question is true. The goal of an argument, then, is to raise this probability value. After learning certain premises (the 'evidence'), we should be more convinced that the theory is true (if the argument is a good one). That is, a good argument raises the probability of the theory or hypothesis in question. We also say that the evidence (i.e. the set of premises of the argument) *confirms* the theory or hypothesis (the conclusion of the argument).

So, how can we model the NAA in the Bayesian framework? The first and very important step involves the identification of the relevant variables. We have to decide which variables matter, and which variables we can safely leave out for the question we are interested in. In the present example, there are two obvious propositional variables: one represents the hypothesis, i.e. that string theory is true (or false). The other represents the somewhat peculiar evidence that the scientific community has not yet found an alternative to string theory. These are two variables, and what we want to show is that they are correlated, i.e. we want to show that the evidence supports or confirms the hypothesis. But how could this be? The evidence in question is certainly quite strange. 'Normal' pieces of evidence are typically deductive or inductive consequences of the tested hypothesis.

An observed black raven confirms the hypothesis that all ravens are black, and the fact that Paul, who is known to be a smoker, has a heart condition confirms the hypothesis that smoking causes heart diseases. The evidence in the NAA does not follow (deductively or inductively) from the truth of string theory. Hence, there is no *direct* connection between the two variables. This makes us wonder whether there is an indirect connection, i.e. that the theory and the evidence are probabilistically correlated, mediated by some other variable. This is something our modelling framework suggests. More specifically, what we have to do is to find a 'common cause' variable that is the cause of both variables and establishes the correlation between them. Such variables are well known in the theory of probabilistic causality, there is Reichenbach's famous principle of the common cause, and the whole theory of 'causal discovery' in Artificial Intelligence is based on it.

Just a question regarding this point. So, to identify this intermediate or third variable, is it using your experience having looked at similar problems in another setting, how does it work?

One always has certain paradigmatic examples in mind. So, for this common cause situation, my favourite example is this: yellow fingers and a heart condition happen to be positively correlated: If you see someone with yellow fingers, you consider it to be more likely that this person has a heart condition than some arbitrary person in the street. And yet, the yellow fingers are not the *cause* of the heart condition, and the heart condition is not the cause of the yellow fingers. For example, you cannot get rid of the heart condition by painting your fingers green or so. The reason for the correlation of the two variables is rather that both have a common cause, viz. smoking, which is the cause of the yellow fingers and the cause of the heart condition. And it is this common cause that correlates the two other variables. But one can say even more: if the common cause variable is instantiated, i.e. if we consider only the subgroup of smokers or the subgroup of non-smokers, then the two variables (yellow fingers and heart condition) are uncorrelated. This makes sense: If we know that someone is a smoker, then learning about the person's yellow fingers does not give us any new information about the probability that the person has a heart condition because the cause of the heart condition is the fact that the person smokes, and this we know already. However, if we do not know that a person smokes, then the observation of the yellow fingers makes it more likely that the person is a smoker (it confirms that the person smokes), which in turn makes it more likely (or confirms) that the person has a heart condition. In any case, this is a nice example that illustrates the idea of a common cause. It is a paradigmatic example that we can also use in other contexts, which I find quite helpful.

From now on, I will relate yellow fingers and string theory!

Yes, the million-dollar question then is this: what is the common cause variable in the case of the NAA? Which variable could do the job here? Well, we argue that the common cause variable we are looking for specifies the number of alternatives to string theory that also satisfy the conditions string theory satisfies. So, we assume that scientists entertain beliefs about the number of alternative theories, that is, if you couldn't find an alternative theory despite a lot of effort and although you consider yourself (and indeed your whole scientific community) to be very smart, then you probably think there can't be that many alternative theories, perhaps three or four, maybe five, but in any case not that many. If there were many of them, then you would have found at least one of them already. It is therefore plausible that the number of alternatives is the variable we are looking for. But does it satisfy the common cause condition? This needs to be checked. So, let's keep the number of alternative theories fixed, let's assume we know that there are 10 of them. String theory is one of them, and they are all equally good (in the light of the available evidence, etc.). What, then, is the probability that string theory is true? Well, you might argue that given that there are 10 alternatives that are all equally good, the probability that string theory is true is simply 1/10. After all, there is nothing special about string theory (apart from the fact that we found it already), and so it seems reasonable to assume that all 10 theories are equally likely to be true. Let us now assume that we find out that none of the nine other theories were found. Would this change our assessment of the probability of string theory? No, because all that matters for this is that there are nine alternatives. Whether we found one or more of them already is of no relevance for the assessment of the probability of string theory. At the same time, if we are uncertain about the number of alternatives and if we cannot find one, then we increase the probability that there are only a few alternatives, which in turn raises the probability that the one we have found is true. Hence, the number of alternatives is the common cause variable we were looking for.

Identifying this third variable and arguing that it is the common cause of the other two variables is the crucial part of the model. It turns out that what follows is absolutely straightforward. To complete the model, one has to make a number of other assumptions about the prior probability distribution over the common cause variable and the likelihood of the other two variables (given the values of the common cause variable); these assignments are natural and uncontroversial. This, then, completes the model and one can use the probabilistic machinery (i.e. the theory of Bayesian networks) to calculate the probability that string theory is true given that no alternative has been found so far. It is then a purely mathematical question to explore under which conditions this probability is larger than the prior probability of string theory, i.e. the probability we assign to the truth of string theory *before* we learned that no alternative has been found so far. We then find the probability of string theory does indeed go up and that the NAA is a good argument.

But wait. Perhaps we should be more careful and check our assumptions again. Given what I said about the common cause variable, it is clear that there is no confirmation if we are certain about the number of alternative theories. The NAA gets its power and strength from the fact that we are uncertain about how many alternative theories there are. If we knew this, then there would be no confirmation and the NAA would not go through (just as yellow fingers do not confirm heart disease if we know that the person in question is a smoker). Hence, to get confirmation, i.e. to make the NAA work, we have to argue that the number of alternatives is not certain. This is where philosophy of science enters the scene. A well-known thesis from this field is the so-called underdetermination thesis that states that scientific theories are underdetermined by empirical data. For a given (finite) set of data, there are always infinitely many curves which go through these data, and hence infinitely many theories that imply the data. This thesis suggests that there are always infinitely many alternatives to a given theory, that is that we should be certain that there are infinitely many alternatives, and this threatens the NAA. Hence, our model of the NAA shows that a defender of the NAA has to show that underdetermination is somehow restricted (at least in the case considered). It has to be shown that it is at least plausible or conceivable that the number of alternatives is finite, and my collaborator Richard Dawid has given reasons for this in his terrific book *String Theory and the Scientific Method* (Cambridge University Press, 2014).

So, what I have just sketched is the model of an argument type. We have seen that modelling the NAA helps us evaluate the argument, and it helps us see which claims have to be further substantiated if we want the NAA to succeed. This can be done with other argument types as well, and I do indeed believe that it is an important task of the philosopher of reasoning and argumentation to identify argument types and to critically assess them by modelling the argument type in a normative modelling framework (such as Bayesianism). Does this make sense?

That's very clear. Just to make sure I am following really well: would you say that a key step in modelling in your field is identification? Identification of the questions you are looking at, identification of the factors, identification of the relationship between the factors — is that a good representation?

Yes. We first have to get clear about what the precise question is we are addressing. Then we have to identify a modelling framework, which imposes a number of constraints on the further assumptions we have to make. Then we identify the relevant variables (and forget about many others) and specify the relations that hold between them. In a probabilistic framework like Bayesianism, these assumptions typically involve assumptions about conditional probabilistic independencies (such as the common cause structure in the NAA). Once all this is done, we can use the mathematical machinery to draw conclusions from the assumptions we made and compare them with data (e.g. about the judgements of scientists) or our intuitions.

So, now a sub-question about the modelling process itself. What is the role of mathematics in modelling?

That's a very interesting question, and I guess other people who are more "philosophical" than I am have much more to say about it. I tend to think that mathematics is just a useful tool which forces us to be precise, to make all assumptions explicit, and which allows us to draw conclusions in an automated way from a set of assumptions. It turns out that the mathematical machinery is enormously powerful. Often, I would not have been able to arrive at the conclusions otherwise. Things are often simply too complicated. How, for example, could we have analysed the NAA without the use of mathematics? Using mathematics allows us to consider more complicated scenarios, scenarios involving several variables, complex dependencies, etc. In my view, the use of mathematics and mathematical models in particular does and will lead to much more progress in philosophy. Often, philosophers focus on highly idealised scenarios, hoping that these scenarios somehow inform us about our rather complex world. Relaxing some of these idealisations and exploring what then follows is often simply too complicated. However, if these assumptions are integrated in a mathematical model, and if one then derives conclusions from them, then many new and important insights and results can be obtained. Or so I believe.

All this stresses the practical role of mathematics. It's convenient and powerful; it forces us to be precise, and it is just a lot of fun to work with.

And what constituents besides this sort of mathematical toolbox or these mathematical formalisms are part of the model? What are the key things?

There is, of course, the modelling framework and there are the assumptions. Perhaps the reasons for the assumptions are also part of the model. I am not sure about that, but it is clear that the assumptions need to be justified and that the conclusions of the model are only as good as the assumptions. 'Garbage in, garbage out,' as they say. The reasons for the assumptions are often rather informal and are not stated in precise mathematical terms. This is completely fine. Another part of a model, especially of the simplified toy models that I like to use in philosophy, are visualisations. When I work on a model, I use all sorts of diagrams, indicating connections between variables, etc. This is especially useful when I develop a model and when I teach it or explain it to colleagues.

This links to the next question. What is the role of language in modelling, and are there qualitative aspects in modelling?

Another good question. I think that the language, i.e. the modelling framework, has to be convenient in the first place. It also has to be justified, i.e. we should have good reasons that the framework is appropriate for the problem at hand. But the practical aspects are also very important. The modelling framework should, for example, be as simple as possible. It is always good to start with the simplest framework first, and then move on to more complicated ones, once a problem shows up. The same holds for the model.

Some of the assumptions of a model are qualitative (e.g. assumptions about conditional independencies), and often we are only interested in the qualitative conclusions of a model. Given that many of the assumptions of a model are highly idealised, we cannot expect the model to get all the numbers right. But we can expect it to explain so-called stylised facts, i.e. qualitative features or patterns that abstract the messy details away.

And then the role of language is probably very important, for this qualitative statement?

There are at least three different notions of the term "language" here and they should not be confused. First, there is "language" in the sense of "modelling framework." I have used this notion before, and clearly the modelling framework matters for the conclusions we draw. Second, there is the language we use to informally talk about the model and to present the model. This could be English or German or any other natural language. Clearly, it does not matter which of these languages we use for the conclusions we draw from the model. Perhaps some language is more convenient for a certain purpose because of the specific vocabulary it employs or because of certain structural features it has; perhaps some assumptions are more straightforwardly formulated in one language than in another language, but at the end of the day the language does not make a difference. This is because the model is presented (and this is the third notion of "language" I want to point out) in the language of mathematics. We introduce symbols and use various structures provided by mathematics, and this helps us to solve the problem the model was constructed to solve.

I think that's exactly the point, because in some other subjects, subjects which are more mathematical, somehow when you write down an equation, there is no doubt about what you mean when you write x is equal to y. Apart from the question of notation, which we'll come back to later, when you write a statement in words there is an importance as to how you qualify the different terms you use, and I guess, because it's not a model for yourself only but is also for the community, and an international community, there is probably this question of how you qualify things. And for instance, German is an extremely rich language for qualifying abstract things very precisely, whereas English is not so rich. So, I was wondering whether this plays a role in modelling? Whether this is taken into account in any way?

I think it is important how variables are denoted. A good notation makes it much easier to reason with the model. It is more intuitive. I remember, for example, that some of my colleagues, when I was working in a physics department, wrote computer programs (at the time in FORTRAN) and denoted all variables that showed up in the program with $x1, x2, \ldots$ Here, $x1$ denotes the variable that occurred first in the program, $x2$, second, and so on. This made it really hard to read and understand these programs, at least without using some kind of dictionary... So, I think good notation is important and makes a difference for the use of the model and also, perhaps, for understanding it easily.

Relatedly, we may ask how a problem is presented. In the models I work with, the problem is always presented in ordinary language. Take the NAA as an example. Here the question is simply, 'Look, this is how people reason and argue; are they justified to do so? Is the NAA a good argument?' However, to address the question (and to solve the problem) we have to reformulate the problem and transform it into a precise question that can be addressed with formal means. This reformulation or transformation is a non-trivial step. It requires the choice of a formal framework (Bayesianism in the case of the NAA), which may be non-trivial, and it requires additional assumptions, which are partly suggested by the framework (e.g. that a good argument makes the conclusion more probable). We can then work with this model, do calculations, and obtain certain results. This, however, is not the end yet. To get an answer to our original question, we have to translate the formal results back and make sure that the original question is properly answered. In the case of the NAA, this goes something like this: 'The NAA is a good argument if you can show that the number of alternatives to string theory (or whatever theory you consider) is not known to be infinite (or does not have any other fixed value).' This translation process is more or less straightforward depending, for example, on the question to what extent it is possible not to use the jargon of the model when answering the original question.

Models are often said to represent a target system. Does this describe what happens in your field? If so, how does the model represent the target, or how do you understand the interface between model and world?

What I just said is very similar to modelling in science. If you are interested in calculating the period of swing of a simple gravity pendulum, for example, you represent the pendulum in a certain way by a model, you then work with the model, calculate things within the model, and finally translate your results back and apply them to the pendulum under consideration. In the case of the NAA, the model is the object we reason with. It helps us make the right inferences, it tells us which (perhaps hidden) assumptions need additional justification, and it comes with a normative framework.

We are also working with other models which are even closer to how modelling in science works. For example, we are interested in the emergence of norms, that is the question how certain patterns of behaviour are formed, and also how so-called "bad norms" (which are norms that no one really wants, such as binge drinking among college students) emerge. How is this possible? Is it possible that a bad norm emerges in a group of individually rational people? To address these questions, we have to make idealised assumptions about the various agents and compute the dynamics of the group in a computer simulation. The most famous (and also one of the first) such agent-based models is Schelling's model of segregation (Thomas Schelling, *Micromotives and Macrobehavior*, W.W. Norton, 1978 (revised 2006)). It aims at providing an explanation for the phenomenon of racial segregation in cities like Chicago. Here, we find the white people in one part of the town and the black people in another part of the town. Why? One possibility is that people are racists and do not want to have anything to do with people of the opposite colour. In a simple model, Schelling showed that another explanation is possible. He put black and white people, represented by pennies and dimes, on a grid with a number of empty spots and introduced the rule that in each "round" of a game people move from their spot to one of the empty spots if less than 30% of the people that surround them have the same colour as they have. The idea is that people have a preference to be among people who are similar to them, but it is not necessary that everyone around them has the same colour. If one iterates this game, then we do indeed find that the equilibrium is a segregated state. Interesting! Hence, this model gives us one explanation for the phenomenon of racial segregation. It explains the general "type," but it does not necessarily explain why, say, Chicago is a segregated city. So, the model does not directly represent and explain a specific phenomenon, but it helps us to reason about certain phenomena in the world and it provides us with how-possibly explanations of them. And this is already something.

The Schelling model illustrates that there is a continuum of models used by social scientists and philosophers. There are also many resemblances to models in, say, physics, where highly idealised models (often called "toy models") are also very popular and serve similar purposes as models in the social sciences or philosophy.

It leads very well on to this question: what is the aim or use of the model? For example, learning, exploration, optimisation, exploitation. So, this is really about exploring what the possibilities could be?

There are many different types of models, and the various models have many different purposes (including the ones you just mentioned). There are toy models that are mostly used to explain and to provide understanding, and there are also very complex and data-intensive models. This makes it hard to say something in general about models.

I am personally most interested in toy models. I like it a lot when a very simple model provides us with some insight. But we are also working with data-intensive models, and sometimes the toy models and the data-intensive models connect. For example, one of our projects focuses on a particular kind of a bad norm — the phenomenon of bullying (e.g. in schools). Here we construct, on the one hand, toy models that aim at explaining, in simple terms, how the phenomenon comes about, and on the other hand, we do network analysis on the basis of empirical data and hope that, at the end, the toy models and the data-intensive models somehow connect. More specifically, we would like to construct toy models explaining new and perhaps surprising phenomena suggested by the network analysis.

We mentioned this earlier: what is the relationship between models and the theory in your field?

I like to distinguish between *models of a theory* and *phenomenological models*. Models of a theory are models that are constructed within the framework of a well-established theory. The already-mentioned model of a pendulum is a case in point. It is constructed in the framework of Newtonian Mechanics. The model of the NAA is another example. It is constructed in the framework of Bayesianism. Here, the framework constrains the modelling assumptions, and — in the case of models of reasoning and argumentation, for example — provides normative guidelines that we need to assess new types of arguments. Phenomenological models are models that are independent of a theory or framework. They are constructed to account for a certain phenomenon (or class of phenomena). Sometimes some of the assumptions of the model are inspired by a theory (or several theories), and sometimes not. The Liquid Drop Model of the atomic nucleus is a good example. Other examples include the Schelling model or the various models for the emergence of norms we construct.

An interesting question to ask about models of a theory is how the framework of the theory is justified or confirmed. Can it be confirmed at all, or is it simply a matter of choice and what we confirm are only the specific modelling assumptions? One advantage of a theoretical framework is that it is used for many different applications. Different models are constructed within the same framework, and this — in turn — gives some credence to the specific models. Or so one may hope. This way of justification does not apply to phenomenological models as there is no theory in the background. These models can only be justified 'locally,' based on how well they explain the phenomenon they aim at accounting for (or whatever the purpose of the model is).

In case you use computer simulations, what would be the relationship between simulations and the model?

The computer simulations we conduct are all based on a model. The model comes first, and then we implement the model on a computer and let it run. This implementation process may involve some non-trivial steps, but the idea is that there is a strict separation between the model and the simulation. Many models can only be studied with the help of a simulation. This also applies to very simple models such as the Schelling's model. (It is true that Schelling himself used pennies and dimes, but today no one does this anymore...) This is why we teach the students in my institute, the Munich Center for Mathematical Philosophy (MCMP), how to use computer simulation software such as *NetLogo*. It is important that students know how to code and how to derive consequences from their models using simulations.

So would you say it's a way to achieve the conclusion in a quicker way? Or is it part of the model in the sense that you might have some back and forth, so you can test something with simulations and you obtain a conclusion, then you come back maybe to your initial phase?

The process of modelling and simulation is certainly much more dynamic than I just said. There is a lot of back and forth between the model and the simulation. Based on the outcomes of the simulation, one introduces changes to the model, and so on. But we do all this also when we do analytical work. You start with a model, you solve the equations by hand and get a certain result. Then, depending on how you assess the result, you make changes to your model. So, there does not seem to be anything specifically new about the methodology of computer simulations in this respect.

What is a good model?

That's a great question and I thought already several times about organising a workshop or editing a book on it. There is probably no easy answer to it as there are so many different types of models. All models have specific purposes, and a good model is one that does well with respect to this purpose. Take weather models. These are extremely complex models, and all we want from them is reliable predictions for the weather in the next days. If a weather model does this, it is a good model, and if not, not. Toy models, on the other hand, have completely different purposes and need to be evaluated on different grounds…

So, what is a good toy model for you?

A good toy model starts with very simple and innocent assumptions and yields a surprising result. A result we did not expect. Schelling's model is a case in point. It leads to the surprising result that the phenomenon of segregation is obtained even if the preferences of the people are not at all racist. Toy models also provide us with understanding. They give us a simple picture or story that makes plausible to us how and why the phenomenon in question came about. Here the model does not provide us with a full story, but a good toy model identifies the main causal factors, the main variables, and strips off everything that is not really relevant. Good models provide us with these insights, and this is what we often want from science.

It is not easy to say more precisely what it means that a toy model provides understanding. Understanding does not seem to be a completely objective notion. An expert may well have a different kind of understanding than a student, and yet, it is important to attempt a more precise explication of what it means that a model provides us with understanding. Several philosophers of science have written on this, and two colleagues and I are currently also trying to make some progress on this thorny problem.

Very good. So, last question on modelling: what has been the impact of the development of new technologies or tools in your field? For example, in cosmology the development of telescopes has been a massive thing; or computer simulation for DNA, that kind of thing. So, in your field, has this had any impact?

Personally, I prefer models which I can construct on a sheet of paper and solve analytically. It gives me a lot of pleasure to play around with a model and to analyse it. And above all, I can do this independently of any technological innovations. But I occasionally also use software such as *Mathematica* for more complex computations and for preparing plots for presentations and publications. I am also pleased that there is software such as *Netlogo* for running agent-based models. This is a fantastic (and even free!) simulation environment that is quite flexible and easy to learn. It is so easy to learn that we teach it to our master students (who typically do not have a background in programming) in one semester and at the end they come up with their own models and simulation experiments. In my view, modelling and simulation are powerful styles of reasoning and thinking, also in philosophy, and it is important that our students learn early on how to use them properly.

Let's move on to the second phase. These are questions regarding uncertainty and risk. How would you define uncertainty? And how does the model help us understand uncertainty in your field?

Uncertainty is a very interesting and important concept and there are several theories around that aim at explicating it. The easiest one is probability theory, and this is the theory I mostly use in my own work. Here, uncertainty means low probability. The lower the probability of something, the more uncertain we are about it. But there are other ways of explicating uncertainty. For example, we may not even be in the position to assign a probability. Perhaps we are only prepared to say that the probability is in a certain interval. If this interval is the closed interval from 0 to 1, then we are maximally uncertain about the event in question. (Theories like Dempster–Shafer or the theory of imprecise probabilities have much more to say about all this.) Here, we have presupposed that there is a probability, but that we do not or cannot know it. There may also be events for which there do not exist any probability. It is interesting to speculate about this, but in any case, the point I want to make is that what we mean by "uncertainty" is relative to some theory or model such as probability theory. Which of them we use depends on the specific application we are interested in. For the problems and questions I work on, probability theory often suffices, but I am getting more and more excited also about imprecise probabilities (and several of my colleagues in Munich are contributing to this rapidly evolving literature).

So, the last question, which is a more personal question. For you, what is the experience or result in your field that has had the most significant impact, and why? It can be your own research, a paper or something, or something more general.

I think that there is not one single experience or result that has had an impact. It is more the fact that mathematical and computational methods are slowly but steadily becoming part of the mainstream in philosophy. The situation was very different when I started my career in the mid-1990s. At the time, the group of formal philosophers was quite separated from the rest and they often addressed different questions others did not care much about. This changed. Meanwhile, formal or, as we like to say in Munich, mathematical philosophers address all sorts of philosophical problems, they collaborate with non-formal philosophers, and their work is quoted in normal philosophical papers. This is wonderful, and it is the result of the openness of mathematical philosophers to use all sorts of mathematical and computational methods and tools that one needs to solve the problem at hand. The next big challenge will be to systematically integrate empirical data in our philosophical models. To do so, we will surely collaborate more closely with experimental philosophers, and I look forward to participating in these endeavours.

CPSIA information can be obtained
at www.ICGtesting.com
Printed in the USA
BVHW010420130520
578889BV00001B/1